DU PONT

E. I. dubont de Nemours

DU PONT

One Hundred

and

Forty Years

BY

WILLIAM S. DUTTON

★

NEW YORK

CHARLES SCRIBNER'S SONS

1942

⋆ *Foreword* ⋆

THIS is a story of men and goods, of peace and war, of vision and venture. It is the biography of tens of thousands—named and unnamed, famous and obscure—whose efforts and identities are merged in the record and character of a company unique in American industrial history.

In 1802, Eleuthère Irénée du Pont de Nemours, refugee from a French dictatorship, established near Wilmington, Delaware, a factory for the manufacture of gunpowder. The latter was more than a necessity in the United States of 1802. It was a means to life and to the national growth.

Gunpowder helped blaze the way for each new frontier as the course of empire swung westward. It helped provide food, raiment, to clear the way for the plow, and later for the locomotive. And it helped fight America's wars.

Dynamite was invented, the most powerful working-tool yet placed in the hands of man. Irénée du Pont's descendants, carrying on the business in unbroken line, pioneered in introducing dynamite to the roaring America of the "Eighties." Quickened progress in the fabrication of steel, the building of railroads, skyscrapers and highways soon made the United States the Wonderland of the nations.

The World War of 1914–18 saw the Du Ponts supplying one-and-a-half billion pounds of explosives to the Allied armies, a prodigious shipment which included 40 per cent of the smokeless powder fired by Allied guns. That industrial feat probably was without parallel.

The Du Ponts had come far since 1802, but in a postwar metamorphosis they were to go farther. They were to put science in overalls and harness the laboratory with the factory on a scale comparable with the American triumph of mechanical mass production.

Today, the Du Pont name is attached to thousands of chemical products, many of which are wholly new to man or nature, and which collectively are effecting a revolution more significant to mankind and its future than any ever promoted by drums and gunpowder. From such common materials as coal, water, air, limestone, salt, vegetable oils, cotton and wood, the Du Pont laboratories are synthesizing the building blocks of a new industrial age.

Our literature is rich in the biographies of individuals who have contributed to the American advance. Statesmen, soldiers, preachers and explorers, poets, inventors and Indian chiefs parade in a glamorous procession across our library shelves, most of them under the ægis of admiring authors. On the other hand, there is a dearth of what might be called the biographies of business organizations.

In this silence, there cannot but be lack of understanding. Such lack would be of small consequence if we still were predominantly an agricultural nation, or if our progress depended chiefly upon the arts or the development of some new political system. However, the inescapable fact is that we are predominantly an industrial nation. We excel mainly because of the efficiency of our mines and manufactures, because of our machine-tilled fields and our machine-made goods and the leadership of our industrial laboratories. Even our agriculture has become essentially a mechanized industry, which looks hopefully to the factory as its potentially largest customer.

It is, therefore, more than desirable, it is vital that we Americans should understand something of American business. This book is an effort toward a better understanding of one American corporation.

I wish especially to acknowledge my indebtedness to the writings and translations of Mrs. B. G. du Pont, upon which I have drawn liberally for material dealing with the period prior to 1900, and for the patient assistance that has been given me by my numerous collaborators in the Du Pont Company.

More than fifty executives of the company, including members of the Du Pont family, have contributed in the form of suggestions. The result is the Du Pont Company as seen by Du Pont men. It is an "inside" view. If that view happens to be good it is only because for 140 years Du Pont men—and women—have striven earnestly to make it so, each generation according to the concept of its time.

W. S. D.

★ Contents ★

CONTENTS

BOOK THREE
A NEW CENTURY

BOOK FOUR
THE PLOWSHARES

[x]

*

BOOK ONE

*

Gunpowder

"No privilege exists that is not
inseparably bound to a duty."

Pierre Samuel du Pont de Nemours

Sword of Honor

SIXTY miles south of Paris, just outside the red-roofed village of Chevannes in the Nemours district, was situated in 1784 the country seat of Pierre Samuel du Pont, Esquire, the Inspector General of Commerce in the cabinet of His Majesty, King Louis XVI of France. Born 1739 in Paris, the son of Samuel Dupont,* a watchmaker, and of Anne de Montchanin Dupont, daughter of an impoverished noble family of Burgundy, the Sieur du Pont had acquired this charming property only in fairly recent years. It was an evidence both of his tastes and of the position to which he had attained in the nation's political and literary circles.

The place made no pretensions that might keep people at a distance. On the contrary, it invited. Every room had a spare bed. The main house of plaster-coated stone walls and steep tile roof was joined immediately by the carriage house, the stables and the barn with their warm odors of cattle and horses, the cackling of poultry and all the patter of farm life. Meadows impinged upon the gardens and

---

*Samuel Dupont used this form of the family name, but beginning in 1763 his son signed himself "Du Pont." Later, he added "de Nemours" to his name to prevent confusion with two other Duponts in the French Chamber of Deputies. Du Pont, in English, is pronounced with the accent on the second syllable. In French, neither syllable is accented.

lawn. A wood at the back was crossed by ditches, which, according to tradition, were the remains of Roman trenches, hence the name of the modest estate—Bois-des-Fossés—meaning literally "Wood-of-the-Trenches."

At forty-five years of age, the Sieur du Pont was a medium-size man of graying and thinning hair, enthusiastic blue eyes, square face and cleft chin. A lifetime devoted to the public service as a writer, an editor and a leader in liberal political thought had brought him many distinctions but not riches. In addition to his post in the cabinet, he was a member of seven Academies, an Honor Councillor of the King and the Republic of Poland in the Council of National Education, Conseiller Intime des Légations to His Serene Highness the reigning Margrave of Baden, a Chevalier of the Royal Swedish Order of Vasa. Within the year he had been elevated to the French nobility in recognition of his services in helping to effect the Peace Treaty of Paris, by which England had recognized the independence of the new United States and ended the American Revolutionary War.

Life, however, is a thing of contrasts. This, the year that had brought to Pierre Samuel du Pont his greatest happiness, his first real freedom from debt and as yet his highest honor, also had brought his greatest loss. The day was the seventh of September. Three days before, in the churchyard of Saint Sulpice at Chevannes, he had buried his wife, and his two sons had buried their mother, Nicole Charlotte Marie Louise le Dée.

She died almost without warning. Late in August, the

weather had turned chill and she had complained of a headache and fatigue. But, she had written the Sieur du Pont in Paris, she would be all right "tomorrow."

It had been her last letter, the last of one thousand and thirty-five letters that Marie Louise le Dée had written her husband during their courtship and marriage.

Now it was afternoon. A hush hung over Bois-des-Fossés. In the living-room, with its great fireplace and blackened beams, the candles had been lighted, although the sun still gleamed upon the windows. The candles flickered over a strange and feudal scene. Alone, seated in a huge armchair and dressed as for a court ceremonial, was the Sieur du Pont. His face was bare of emotion. His sword, embossed in gold with the Du Pont coat of arms and motto *Rectitudine Sto* (By Uprightness I Stand), glittered at his side. At his feet was a silken cushion, to his right an empty chair in which lay the sheathed sword and cocked hat of a young man. On his left stood the bust of his dead wife, white as alabaster.

A manservant entered the room. At a nod from his master, he proceeded to draw the curtains over each of the windows until he had shut out the last of the day. Shadows now masked most of the familiar objects of living —the pianoforte, the table for backgammon, the comfortable straw-bottomed chairs. In contrast, the Sieur du Pont, erect in his armchair, sat as upon a lighted stage.

Again he nodded in a prearranged signal, as all that was to follow had been prearranged and set down by him in writing to be preserved by his descendants. Through a doorway appeared his two sons. The elder, Victor Marie, strode forward leading his brother by the hand. Victor, too, wore his sword but he carried his hat in token of respect to his sire. Seventeen, handsome, a young giant of six feet three-and-a-third inches, he spoke:

"My father, I bring my brother to receive your blessing and his first arms."

"Place him before me," bade the Sieur du Pont, and arose.

The younger son stepped into the circle of candlelight. His boyish face was as emotionless as his father's. Born in Paris on June 24th, 1771, Eleuthère Irénée du Pont had entered his fourteenth year. The time had come for him to be instructed in the responsibilities and duties of manhood, and in the family's code.

The Sieur du Pont's face softened as he looked upon his younger son. Blue-eyed, shy, quiet, practical, sometimes hard to understand, Irénée was like his dead mother.

"My son," he said, "I had hoped that your mother's presence would honor the ceremony in which I endow you with your rank. I expected to arrange a fête which she would have adorned and have shared with us. There are no more fêtes for us, my son. Nothing is left but to exercise our courage and to fulfill our duties.

"You saw me gird on your brother's sword. Your birth gives you a right equal to his. You must understand, I have told you often, that no privilege exists that is not inseparably bound to a duty. The Nation confides the privilege of bearing arms to such families as she deems most distinguished, but this privilege cannot permit the arbitrary or careless use of arms for oppression or for harm. It involves, on the contrary, a special obligation to protect innocence and weakness, to oppose injustice, to maintain peace, to devote itself to the service of mankind."

The Sieur du Pont resumed his seat in the armchair.

"Come here, kneel."

The younger Du Pont knelt on the cushion.

"Put your hands between mine."

The boy raised his hands and his father clasped them between his own.

"Promise that you will never give way to anger and still less to hatred to such a degree that you will shed the blood of any man unless you are constrained by the most absolute necessity. But promise at the same time that you will not allow yourself to be cowed by any danger when you are called upon to defend your country, or your wife, or your children, or your brother, or yourself, or any other human being who, in danger not deserved by his own wrong-doing, has need of your help."

"My father, I promise it to God," said the kneeling boy, "to my country, to mankind, to the memory of my mother and to you."

Taking the sword from the empty chair the Sieur du Pont girded it about his son. He stood and drew his own sword. With its flat side he struck the boy on the left shoulder.

"The blow that I am giving you, my son, is to teach you that you must bear any blow when it is honorable and right to accept it. Rise and embrace me!"

The Sieur du Pont kissed Irénée on both cheeks, whereupon he turned to Victor. "My son, give me your brother's hat."

The hat was handed to Du Pont *père* and he placed it on Irénée's head. Victor donned his own hat.

"Draw your swords, my children," bade the Sieur du Pont. "Salute and embrace each other!"

The three swords flashed. The brothers embraced.

"Promise each other to be always firmly united, to comfort each other in every sorrow, to help each other in all efforts, to stand by each other in all difficulty and danger. Remember that you are doubly brothers. Others are so only by blood. You are brothers by blood and by honor, by birth and by arms. The same arms and the same motto will remind you always of the same responsibilities. In every way you are bound together by a chain that should be lasting and easy to wear."

The father sheathed his sword.

"I bless you, my children, may Heaven bless you!" he said. "May your works and your children be blessed! May your families be perpetuated by wives who are good, reasonable, brave, economical, generous, simple and modest like the mother you have lost. May each generation of your descendants strive unceasingly to make the next generation better than his own. I shall try to do my best for you; try to do still better for your children."

He bared and bowed his head. For a moment he stood so in the utter silence. Then slowly he walked from the room and Victor followed. Irénée du Pont stood alone with his new sword and new hat, a slight, awkward figure that suddenly had taken on the estate of a man. In the still candlelighted place, his shadow stretched out behind him.

Somehow, sometime, through the decades ahead, work done in that shadow and in the name of this Du Pont was to affect, in small way or large, the daily living of a nation; because Eleuthère Irénée du Pont was destined to found both an institution and a tradition, which, through his sons and his sons' sons and their descendants, would live beyond his century into our own.

The tradition—the code of the Du Ponts—began on that September day of 1784.

Reign of Terror

AFTER his mother's death, Irénée du Pont went to Paris to live with his father and brother in an apartment in the Rue de la Corderie. He did not like Paris. He missed the fields and woods of Bois-des-Fossés, the long tramps with his mother, the garden with its never-ending mystery of growing things.

In Paris, save for his tutor and the servants, he was alone much of the time. The Sieur du Pont was occupied with the affairs of the Department of Commerce, where Victor, too, was employed. Victor often traveled to distant cities to report on their commerce. Occasionally he wrote brilliant letters, gay and carefree, in which he enclosed rare seeds he had come across. These Irénée would plant in little clay pots on the window-sills.

At times, distinguished visitors came to the apartment. They sat by the hour talking of wars, politics, the inconsistencies of the King, and the storm of trouble rising in France. The first faint thunders of revolution were rumbling over the nation. Among these visitors were Benjamin Franklin and Thomas Jefferson, the Marquis de La Fayette, who had fought in the American war, the suave and elegant Talleyrand, the aged Minister of Finance, Vergennes, the English philosopher, Dr. James Hutton, and

Antoine Laurent Lavoisier, the great French scientist and chemist.

More than any other, Lavoisier held Irénée's interest. He was a member, along with the Sieur du Pont, of the *Comité Consultatif d'Agriculture*. They often talked of the soil and crops and the everyday worries of the farm. These things the boy knew and in part at least could understand. Eagerly he listened to their discussions, sitting in a corner so as not to be sent to bed.

One day Lavoisier gave him a paper of salt-like powder and instructed him to mix it with the soil in his flowerpots. Shortly the plants took on new life and Irénée fairly burst with questions. In answer, Lavoisier took him to Essonne, where he was chief of the government gunpowder works, and let the boy watch saltpeter being mixed with charcoal and sulphur to make gunpowder. It was this same saltpeter, he explained, which had invigorated the plants.

Irénée wrote an account of what he had seen and learned at Essonne. The great chemist read it with surprise because of its accuracy. He spoke to Du Pont *père*.

"I need to train a successor," he said, "a young man to carry on my work when I must stop. Perhaps—who knows —he may exist in your son."

Two years later Irénée went to work at Essonne. The making of gunpowder was to be the first phase of his education in the complex science of chemistry, a science already envisioned by Lavoisier, one of its most brilliant founders, as a power capable of changing the world.

Irénée learned how to refine saltpeter and sulphur, to

combine the two with charcoal baked from willow wood, and to press, grain, and polish the product into the explosive known as black powder, or gunpowder. He learned how to size the grains to suit the various bores of guns, how to operate machinery, how to guard against carelessness in one of the most hazardous works in which men engaged. Careless men did not live long in this business.

Four years went by. Lavoisier agreed to make the now tall, rangy, sandy-headed youth of twenty his first assistant in the laboratory. Elated at the prospect, Irénée married, on November 26th, 1791. His bride, Sophie Madeleine Dalmas, was sixteen. The two planned a place in the country where living would be cheap.

"I can imagine no greater happiness," Irénée wrote to his father, "than to live surrounded by those I love."

That year, 1791, "le déluge," foreseen by Louis XV, was beginning.

Already the Bastille had been stormed by a Paris mob. The government, too late, was making desperate efforts at economy. Irénée's wedding was only a few days old when Lavoisier was transferred from Essonne to the Treasury. One by one, Du Pont *père* was stripped of his salaries and perquisites. He was preaching moderation, while those who now controlled in politics were plotting revolution.

Irénée had no interest in Essonne without Lavoisier as his chief, so he rejoined his father, who, at fifty-two, as plain Citizen du Pont, was starting life anew as a book publisher in Paris. As usual, Victor was away, this time in America serving as secretary without pay to the Comte de Moustier, first French minister to the United States. Victor had set his heart on a diplomatic career.

The new publishing house, launched on borrowed capital and a heavily mortgaged Bois-des-Fossés, planned to publish books of science, memoirs and travels, and to issue from time to time political pamphlets in which the elder Du Pont would air his views on current events. The prospectus was mild. Irénée soon found, however, that the printing of books, and especially political pamphlets, could be as dangerous as making gunpowder. Du Pont *père* believed in moderation, but he was ready to fight if necessary to get it. His goose-quill pen spat fire, though the thunder of trouble rumbled closer.

The King was shut in his palace of the Tuileries, virtually a prisoner. There were riots in the Paris streets, Royalist fugitives crowded the roads out of France. Undaunted and undismayed, Du Pont flayed the mob, and, in pamphlet after pamphlet, demanded the constitutional monarchy instead of anarchy. He organized a company of fifteen, which soon grew to "sixty men of the greatest courage," each sworn to give his life, if need be, to preserve orderly government.

Irénée was a member of this band. It allied itself with the National Guard. His work and sleep were now continuously interrupted by calls to arms to aid in putting down lawlessness. Management of the publishing plant fell almost wholly on his shoulders. He had another very serious concern—Sophie, his young wife, was about to become a mother.

On the tenth of August, the mob attacked the Tuileries. It was met by the King's Swiss Guard, and on the right of the Guard stood Du Pont's company of sixty. The King fled to the protection of the near-by Legislative Assembly and the Court followed him. The slaughter that followed —a red blotch on French history—wiped out the Swiss and left Du Pont's company with only eight survivors, including his son and himself. His quick thinking saved their lives. He ordered his followers, battered and powder-blackened, to simulate a revolutionary patrol squad—they looked "revolutionary" enough in their tattered condition. Amid the confusion, the eight marched off to safety.

"Our coolness, our measured and rather slow step like that of men wearied by the day's work, made us seem to be a regular patrol and we went out with no difficulty," Du Pont wrote later in his memoirs.

The day marked the end of the monarchy and the red dawn of the Reign of Terror.

At once a search was begun for Du Pont. He had been recognized among the King's defenders, although Irénée had escaped notice. Searchers stormed the printing shop and ransacked their apartment in the Rue de la Corderie. Du Pont, however, was safely hidden in the dome of the Observatory, a bare cobwebby place presided over by the astronomer Lalande, an old friend. The fugitive had no bed. Food was smuggled to him at night. Through the musty heat of August he stayed in hiding. Then, on September 2nd, a new furor swept the city as massacre was begun of the Royalist prisoners in the jails. The city gates were opened in the excitement. Disguised as an aged physician wearing a dark eye-shade, Du Pont slipped through and into the country.

Irénée stayed behind. Only three days before, Sophie's baby had been born, a girl whom they named Victorine.

As soon as he could, Irénée moved them to Bois-des-Fossés to stay with his cousins, M. and Mme. Lamotte, and a faithful old retainer, Cœur de Roi. Then, alone, he went back to the printery to repair the damaged presses and carry on the work of the shop. Disease, famine and death stalked in Paris.

January, 1793, the King was guillotined; in October, Marie Antoinette. Before and after them, as they were routed from hiding, passed a seemingly endless file of doomed men, as Brissot, Hébert, Marat and Danton each in turn wrote his name in letters of blood across France.

Irénée was subject to service in the National Guard, and calls came almost daily. Because of his experience at Essonne, he was ordered to equip and manage a saltpeter refinery for the government, now at war on a dozen foreign fronts. He started a newspaper, *Le Républicain,* published Lavoisier's latest books on chemistry, printed bonds and paper money for the new Republic, and thousands of copies of the new Republican Calendar.

The guillotine now took as many victims in a day as it had in a week, but reports persisted that the elder Du Pont had been killed, so the search for him was relaxed. He felt safe to return to Bois-des-Fossés, where, in hiding, he plunged into the writing of one of his most notable literary works, *Philosophie de l'Univers*. Irénée's wife, Sophie, managed the farms. Lavoisier promised a visit.

Spring came and with it Robespierre's edict, "The Republic has no need of scientists!"

On the eighth of May, 1794, the great chemist, known and respected by the world, went to the scaffold. Next, on July 13th, a warrant was issued for the elder Citizen du Pont. Nine days later, racked by gout but defiant, he was lodged in La Force Prison in Paris.

Du Pont organized his fellow prisoners, first into a unit to die fighting rather than be led meekly to death, and, second, into a class in philosophy and political economy before which he lectured daily. Dressed as a peasant, Sophie took him food.

But France was wearying of bloodshed. Robespierre himself was arrested. He was executed with twenty-one of his followers on July 29th, 1794. The elder Du Pont and other political prisoners were freed.

Irénée, now twenty-three years old, looked ten years older. His face was gaunt. However, despite the revolution, he had more than doubled the Du Pont printing business. He had founded a successful newspaper. Single-handed, he had supported the family on his earnings.

The Great Adventure

AT THE height of the French Revolution, when every person of birth slept in dread, "a very tall and very handsome young man dressed in the English fashion, whose manner and bearing, entirely unaffected and almost careless, had, nevertheless, a very charming ease," arrived in Paris from the United States.

Thus wrote Mademoiselle Gabrielle Joséphine de la Fite de Pelleport, daughter of the Marquis of that name, in her *Souvenirs*. At the time she, too, was a charming person, of twenty-three. The subject of her pen was Victor du Pont.

Four years of diplomatic service in New York and Philadelphia, in turn the American capital, as a member of the suite of the French minister, had given Victor a poise that promised much for his future in that service. In the France of 1793 and 1794, however, he invited trouble, and it was with mingled pleasure and relief that Du Pont *père,* in April, 1794, saw his elder son and Joséphine de Pelleport married.

Shortly, too, a new American post was offered Victor, this time that of Consul at the City of Charleston. He accepted it, for he liked America and spoke English fluently. Charleston's culture and society were equal to Philadelphia's, and in Charleston Victor du Pont and his wife were to spend the next three years most agreeably.

Seeing them off, Irénée felt no urge to accompany them. He was "Citizen du Pont, Printer," who had made a place for himself in trouble-torn Paris. He did not mix in politics, kept to his own affairs, was respected and liked. Recently, the General Assembly had commissioned him to make collections for the poor—the city was filled with the starving. He was one of a commission named to investigate and correct abuses in the prisons. His printing-presses were busy day and night.

Du Pont *père* returned to Bois-des-Fossés and his writing. Irénée spent eight days there, walking with Sophie over the old byways and helping her with the work. She managed the farms, saw to the planting of the crops, paid the men weekly, made butter, attended to the vintage. Later, from Paris again, Paris still hungry, still sullen, Irénée wrote to her:

"It is said that one grows accustomed to anything, but your absence is more and more unendurable to me. I can never accustom myself to living without you. My happiness and my life depend on our union." The letter ended: "It is eleven o'clock and I am on guard at the National Convention."

Again there were riots. Irénée wrote Sophie on May 26th, 1795: "I have been on duty for five days and five nights, and for thirty-six consecutive hours with nothing to eat or drink."

However, the conservative elements in the nation had

gained strength. A new government was formed under five Directors. The legislature was divided into two Councils. The people of the Nemours district dared to elect Du Pont *père* to represent them in the upper house, the Council of Ancients. New life surged through the veteran fighter. In 1795 he launched a new newspaper, *L'Historien,* and now some of France's most brilliant writers rallied to his standard. *L'Historien* became the mouthpiece of the conservative party.

A new general, Bonaparte, led a French army to sensational victories in Italy. Money and booty poured into the French treasury. The cries for food ceased. Every appearance of the Deputy du Pont on the floor of the Council of Ancients brought applause from the galleries as he lashed again and again at the Directory, demanding an end of dictatorship. An election was held and the conservatives won. Du Pont was elected President of the Council of Ancients.

Victor, in America, was yet to learn of his father's election, when the five Directors defied the electorate and struck back. Du Pont and fifty-three Deputies were seized and thrown into La Force prison. Irénée was arrested. *L'Historien* was suppressed. The police ransacked the printing house and destroyed the presses.

Powerful friends rose to Du Pont's defense, and, after one night in prison, he and Irénée were released. Nevertheless, Du Pont was forced to resign from the Council of Ancients, and police surveillance followed him home to Bois-des-Fossés. One after another, his friends in the

Council of Ancients were deported to French Guiana, the "Devil's Island" prison off the South American coast. The family met at Bois-des-Fossés to consider what should be done.

It was autumn again. Sophie held in her arms her newest baby, Evelina Gabrielle, almost six months old. Du Pont's hair was almost white, his forehead bald, and Sophie noticed for the first time that his shoulders had begun to stoop. However, neither misfortune nor years had succeeded—or ever did—in dampening Du Pont's spirit. At the age of fifty-five, he had just married the widow of his old friend Pierre Poivre.

The fire crackled in the open hearth and above it Du Pont's voice rose and fell. Only one haven, he said, was left them—the United States.

They stared at him. Irénée thought of the years he had worked to establish the publishing house. Sophie saw the trim, freshly plowed fields of Bois-des-Fossés, the vineyards picked clean of grapes, the filled mows and bins in the barn, all work to which she had attended.

"America," repeated Du Pont *père,* "where liberty, safety and independence really exist!"

Already a plan was forming in his mind. By spring, the plan was a daring new adventure. Du Pont proposed to form a company to purchase lands in America and colonize them. The headquarters would be at Alexandria, near the new American capital city of Washington, then being built. The colony itself would be based in the upper and western counties of Virginia "in a beautiful valley above the Shenandoah."

Enthusiastically he elaborated the scheme, poring over maps, working out details, writing into the night, until shortly a prospectus issued from Irénée's remaining presses. It explained how farms could be laid out for the colony, sawmills set up to supply lumber for homes, manufactures for pottery, glass and the like established, and—later— schools built, stores, churches, inns for travelers.

To finance the venture, Du Pont proposed to use all that remained of his own fortune and to raise 4,000,000 francs by sale of stock at 10,000 francs a share. The directors would be himself, Victor, Irénée, and Bureaux de Pusy, son-in-law of the second Madame du Pont. Pusy was an army engineer of reputation and only recently had been released, along with La Fayette, from an Austrian prison on condition that he stay out of France. La Fayette, also denied residence in France, was expected to join the company.

Victor du Pont, the one member of the family familiar with American conditions, knew nothing of his father's plan. He had recently been appointed French Consul General at Philadelphia, but President Adams, at bitter odds with Paris, had refused to recognize him. Victor now returned to France, but too late to convince his father that he was attempting the impossible. Victor warned that foreigners could own land in only a few American states, that the French were unpopular in America, and that the government was actually threatening war with France. Moreover, he argued land prices in the United States were high and going much higher.

Du Pont listened, but his enthusiasm was unquenchable. The thought of war was ridiculous, he said. The great scheme was already under way, some of the farms of Bois-des-Fossés had been sold, Irénée had buyers for the printing business, friends had already subscribed three and a quarter millions of francs for the new company's stock— the stock of Du Pont de Nemours Père, Fils et Compagnie. To his dismay, Victor found himself committed to something from which he could not withdraw. His name and help had been pledged in the company's literature. So had his father's and his brother's. The *family's* word had been given!

They were almost two years in preparation. The delay enabled Irénée to study botany at the *Jardin des Plantes* in Paris—knowledge that would be useful in the new land. In Paris, too, amid the growing confusion of making ready for departure, Irénée's and Sophie's next child, Alfred Victor, was born on April 11th, 1798.

Finally, in May, 1799, Bureaux de Pusy, his small daughter Sara, and his mother-in-law Madame du Pont sailed for America. Pusy was to establish an office in New York and prepare for the coming of the others. One day out, their ship was captured by the English and for six weeks they were held before being permitted to continue their journey.

The main party sailed from Ile de Ré on October 2nd,

1799. It numbered seven adults and six small children, namely, Du Pont *père* with his two sons and their wives; Irénée's three children, Victorine, Evelina and Alfred; Victor's two, Amélie and Charles Irénée, who had been born during the stay in Charleston; Bureaux de Pusy's wife and few-months-old baby, and Sophie's brother, Charles Dalmas, who was twenty-two.

Their ship was the *American Eagle*. Overcrowded, poorly manned and leaking so badly its cargo of salt soon turned to salt water, the battered craft headed westward. Twice the captain lost his course. Twice they ran out of food and had to be supplied by English ships, which answered their flag of distress. The sailors got out of hand and looted the passengers' baggage, so that the Du Pont men had to stand guard at night with swords. Almost everybody was seasick, discouraged and disgusted, excepting Du Pont *père*. He paced the deck, joked at their hardships and wrote poetry for their amusement.

On New Year's day, 1800, after ninety-one days at sea or almost three weeks longer than it had taken Columbus to cross three centuries earlier, they landed at Newport, Rhode Island, which was in the grip of one of its worst New England winters. The ship had been given up for lost and nobody met them.

Half frozen and nearly starved, the party set out to find warmth and something to eat. At the nearest lone house nobody answered Du Pont *père*, although he both knocked and shouted. Finally he peered in a window to see a table loaded with food and set for a meal. A fire crackled in the open hearth. It was too much. Du Pont tried the door, found it unlatched, and led his famished followers in to the feast.

There was some excellent wine, new logs ready for the fire, and they cleaned the table. Still there was no sign of the owner, who evidently had gone with his family to church. They held a conference. Du Pont *père* cheerfully agreed that they may have committed a felony, so he fined himself one gold coin and left it at the head of the ravished table.

Bureaux de Pusy had bought a house at Bergen Point, nine miles from New York City on the New Jersey coast. After another week of winter travel they were all assembled there. Du Pont *père* named the place "Goodstay." The great adventure had begun.

"Project No. 8"

DU PONT'S plan for a Virginia colony never got beyond paper. Everywhere roads, rivers or trails led he found soaring land prices. His old friend, Thomas Jefferson, now Vice-President of the United States, wrote in emphatic warning against investments of any kind without the most careful investigation. Young America was riding a business boom; speculators awaited the unwary.

Since 1775, the population had almost doubled. The new republic had 5,300,000 inhabitants in 1800. Virginia alone had almost 900,000 people, and "west of the mountains" in Kentucky, Tennessee, Ohio and Indiana were a half-million white settlers. What land was to be had at fair prices was mostly in inaccessible wildernesses, to which Du Pont could not picture Frenchmen being persuaded to immigrate.

He soon found, too, that Victor had not exaggerated in saying the French were unpopular in the United States. Indeed, they were only a little less unpopular than the English, because of both nations' offenses against American rights at sea. For that matter, all foreigners were unpopular. The growth in population had been effected with almost no help from Europe, which was involved in seemingly endless wars. The young blood of the land was largely American-born and deeply conscious of the fact. In most states, laws prohibited alien ownership of land, an extra 10 per cent tax was imposed on aliens' imports, and there were stringent requirements for naturalization. Virginia alone granted citizenship upon purchase of real estate, so Victor bought a small property at Alexandria and became naturalized.

Du Pont's greatest difficulty, however, was that he lacked capital. An imposing array of distinguished Frenchmen had subscribed to his colonization scheme, among them La Fayette, Beaumarchais, Duquesnoy, Rousseau, Portalès, La Tour Maubourg, and Necker, but only a few ever paid. Instead of a proposed capital of 4,000,000 francs, Du Pont received only 455,000. Of this, 30,000 francs had been lost in a bank failure before he left Europe.

At "Goodstay" the winter waned into spring, summer passed and autumn turned toward winter again, but still the Du Ponts were without remunerative employment. Du Pont *père* was made a member of the American Philosophical Society. He wrote on natural history for the *Institut National de France* and, at Jefferson's request, prepared a treatise on national education. His fertile imagination saw in this new country infinite possibilities. He conceived plans for a fast mail and passenger service between America and Europe, devised a program by which Spain might elude the vigilance of England, with whom she was at war, and establish reliable communications with Spanish-owned Mexico. He proposed a new company, headed by Victor, to act as intermediary in the exchange of cargoes between France and her colonies in the West Indies and also the Far East. In all, he drew up seven great projects for Du Pont de Nemours Père, Fils et Compagnie. They involved, in the aggregate, millions of francs in potential profits, if they got started.

Irénée fretted in his enforced inactivity. The dwindling of the company's capital worried him; by autumn, a fifth of it had been spent with nothing practical done. He was now in his thirtieth year, and had a wife and three children to support. Twice, through no fault of his own, he had lost what might have become permanent occupations. He could not share his father's optimism.

Then, upon Victor's urging—Victor seemed to have friends everywhere—Irénée went shooting with a French-born American artillery officer, Colonel Louis de Toussard, who had come to America with La Fayette and had remained to make his home here. The Colonel lived on the Lancaster Road near the town of Wilmington, Delaware, which he described as a community made charming by its own little colony of French, who had been driven from the West Indies colony of Santo Domingo about ten years earlier by an uprising of negro slaves.

By afternoon they had shot away their powder, but were able to replenish their supply at a country store. Its high price and poor quality amazed and angered Irénée, whom Lavoisier had taught what good powder was. He felt they had been cheated. Toussard assured him the transaction was usual, that all American-made powder was high-priced and poor according to French standards. The only good powder to be had was that imported from England, a fact of serious concern to the American Army, the artilleryman explained.

That night Irénée paced the floor with an idea. Why not establish a gunpowder mill in America!

The rifle was the inseparable companion of the pioneer's ax. Every fireside had its musket and powder-horn hard by; even women were taught to use them. Hunting and trapping represented an industry that was supplying the world with many of its finest furs. William Astor of the American Fur Company, for instance, was shipping pelts as far away as China! Vast areas were to be cleared of stumps and boulders.

Irénée began to canvass the subject. He questioned Toussard about the number and character of American powder-mills. The colonists, he learned, had made good powder during the closing years of the Revolutionary War, but no effort had been made since to keep up with European advances in the art. Lavoisier's research work, which had greatly improved powder, was practically unknown here. Moreover, accidental explosions and the English competition had driven all but a few mills out of business, Toussard said.

With the French-American colonel as guide, Irénée visited the few American mills still in operation. He found them making bad powder at great expense, but nonetheless earning profits. Back at "Goodstay," he wrote:

"To give an idea of the incompetence of these manufacturers we will take as an example that plant* which has the best reputation and is now working for the Government. The Philadelphia merchants who own this manufacture, nine years ago brought to this country as manager of their plant a Batavian workman who makes their powder as he saw it made in his own country—as they have probably made it in that Dutch colony for fifty years. . . .

"They work four mills night and day, while with two mills, working in the daytime only, their output should be a quarter more than it is. They employ sixteen men; they would need but twelve to do as much work with two mills.

"They grain their powder by crushing it in a wooden sieve or a kind of basket so badly arranged that the greater part is reduced to dust—increasing the cost of labor and the loss in manufacture.

"Such competitors should not be formidable to one who, having studied this manufacture for several years in the powder works of the French government when they were directed by M. de Lavoisier, can add to the extensive knowledge of that administration the important modifications which have been in use since the Revolution and which have been caused in the making of powder by the needs of an unprecedented (European) war."

Irénée calculated that a plant comprising one stamping mill and one wheel mill,† operated under peacetime conditions, should produce a minimum of 160,000 pounds of powder annually. This output, he estimated, could be sold for at least $40,000.

The cost of manufacture would be:

120,000 lbs. saltpeter @ 10¢.......................................$12,000.00
20,000 lbs. sulphur @ 2¢... 400.00
20,000 lbs. charcoal @ 1¢... 200.00

*Probably the Frankford, Pennsylvania, mills of William Lane and Stephen Decatur.

†It is doubtful if any wheel mill existed in the United States at this time, though mills with wheels and pans of black marble had been operated in Europe since 1540.

1 head workman @ $1.75	$ 638.75
4 upper workmen @ $1.50	2,190.00
12 workmen @ $1.25	6,387.50
Director's salary	2,000.00
Loss in manufacture and incidentals	4,183.75
Annual repairs to machinery	2,000.00
	$30,000.00

These estimates indicated a possible net of $10,000 a year, although Irénée had fixed all costs and wages at above the actual market, and the selling price on the finished powder at less than the market. Capital requirements he placed at $36,000.

With these facts he went to his father. The venture looked small and unimpressive as compared with the least of the elder Du Pont's international projects. It lacked the smack of romance and great affairs that to Du Pont *père* was both meat and drink. However, he saw in the proposal a feature that Irénée, his head buzzing with figures, apparently had missed. Such a powder factory would be at once a blow to England's commerce in America, and a pillar of defense against France's hated rival should England ever again dare an American war. Preposterous, of course, the old fighter said, was this Federalist pother about America going to war with France.

So, chuckling at the impending discomfiture of the English powdermakers, Du Pont got out his list of projects and added, as Project No. 8, a gunpowder manufactory in America. Jefferson, whom he consulted as to the attitude of the United States Government, encouraged the plan. He suggested that the powder plant be situated near the national capital.

"Goodstay" bustled with activity. It was decided that Victor and Irénée should return to France, the former to enlist support for the various major projects of the company and Irénée for the powder plant. They sailed January 5th, 1801, and one month later landed at Havre, where each went his way.

Irénée readily found aid. His old associates at Essonne had not forgotten him. Napoleon's militant star was in the ascendency and a determined effort was being made to wreck the commerce of England. Consequently, the chiefs of the French Powder Department offered all the technical assistance at their command. It was arranged that Government draftsmen should draw the plans for Irénée's machinery, that the machines should be built at cost in the Government's shops, and, if necessary, trained workmen should be released to form the nucleus of the American plant's crew.

Likewise, Irénée was pledged financial support, notably by Jacques Bidermann, Swiss banker; Adrien Duquesnoy, who had been a fellow prisoner with Du Pont in La Force; and Louis Necker, uncle of Madame de Staël and brother of Jacques Necker, Minister of Finance under Louis XVI. Moreover, enough money was actually paid to enable Irénée to order his machinery, at an approximate cost of $4,000.

In July, 1801, he was back in America with trunks and boxes loaded with surprises for all the family. He brought Paris bonnets and dresses for Sophie and the children, a box of colors for Victorine, dolls and toys for Evelina and Alfred, and packages of seeds and plants for their future home. But what created the most excitement were seven young Spanish merino sheep. This breed of sheep, famous for its long silky fleece, he hoped to introduce in America.

Also Irénée brought back a formidable-looking document—the *"Articles of Incorporation for the Establishment of a Manufacture of Military and Sporting Powder in the United States of America."* These stipulated that the capital of the company should be $36,000, in eighteen shares of $2,000 each. One share each had been subscribed by Bidermann, Duquesnoy and Necker, the parent Du Pont de Nemours company pledged itself to take eleven, and four shares were held open for possible American subscribers.

Citizen E. I. du Pont was named Director of the enterprise at a salary of $1,800 a year and "one-third of the profits or losses, should there be any." The document had been executed in Paris on April 21th, 1801.

The four open shares were soon taken by Archibald McCall, a Philadelphia merchant, and Peter Bauduy of Wilmington. The latter, who had been born Pierre Bauduy de Bellevue, was one of the little French colony that Colonel Toussard had mentioned. Bauduy spoke and wrote English, and was familiar with American business practices. He agreed to give part of his time to the correspondence and sales work of the new company.

All that remained was to select a suitable site for the powder works, with Bauduy presenting strong arguments for locating in Delaware. Irénée offered to buy the Frankford plant of William Lane and Stephen Decatur, but they refused to sell. He explored the country near Washington, the Hudson River Valley in New York State and various sites suggested in New Jersey, but finally, to Bauduy's delight, came back to Wilmington. Four miles west of the Delaware town, on the farm of Squire Jacob Broom along the rock-strewn Brandywine, he found what he sought.

On April 27th, 1802, the Broom farm of ninety-five acres was purchased for $6,740 in the name of William Hamon, a naturalized Frenchman and friend of Peter Bauduy. Delaware law made it impossible for young Du Pont to hold the land in his own right. That was to await his qualification as a citizen of the United States.

Meanwhile, Victor du Pont was getting started in the commission business in New York City. Unable to interest Europe in his father's projects, he had agreed to act as agent for Louis Pichon, French Consul General at Washington, in provisioning French troops in Santo Domingo. The Bureaux de Pusys had returned to France. They and other friends urged the elder Du Pont to do likewise, pointing out that Napoleon was welcoming back the émigrés of the Revolution. Smelling action, Du Pont decided to re-

turn.‡ His two sons were now busy with their own under-
takings. In person, he could better explain to his French
stockholders the failure of their original enterprise and
the promises of the new.

‡On this trip, Du Pont de Nemours also carried confidential letters
from President Jefferson to Chancellor Livingston, the American Minister in Paris, regarding the pending purchase of the Louisiana Territory. *"The Correspondence of Jefferson and Du Pont de Nemours,"*
by Professor Gilbert Chinard, which was published in 1931 as one of
Johns Hopkins University Studies in International Thought, credits
Du Pont with an important rôle in the Louisiana negotiations.

The Mills on "the Brandywine"

IRÉNÉE DU PONT, with his wife and three children, moved to the former Broom farm on Brandywine Creek near Wilmington, on July 19th, 1802. It took them four days to drive the 130-odd miles from Bergen Point, New Jersey, over roads deep with ruts and dust. Their house on "the Brandywine," as the stream was to become affectionately known to them and succeeding generations, was the first of their own that Irénée and Sophie had possessed. It was a crude, two-story structure of native stone built in the hillside. With low ceilings and few windows, the house was uncomfortably hot under the summer sun.*

Sophie, who had dreamed of finishing her days at Bois-des-Fossés, saw roughly plastered walls and bare plank floors, littered with the confusion of a family in transition. They had sent their heavy baggage ahead by boat, the schooner *Betsy of Patterson,* to New Castle, Delaware, from where under the charge of Sophie's young brother, Charles Dalmas, it had been carted eleven miles to the farm. Boxes, barrels and crates sat about, the unassembled parts of beds, pots, skillets and dishes, bundles of bedding and winter cloaks, books, and children's toys.

*This house still stands, in good repair.

Outside were farm implements, garden tools, casks of wine and vinegar, and Irénée's pair of hunting dogs panting in the heat. The merino sheep bleated hungrily from a near-by shed. But it was a home at last!

The ten-plate cookstove had been set up, smoke drifted from the stone chimney, and presently from the kitchen that was also the general living-room came the tempting odors of food being prepared. The Du Ponts were settled in Delaware.

Neither of the two newspapers then published in Wilmington, a town of about 3,500, was concerned enough to mention the fact. Not until January 1st, 1806, did the name of Du Pont appear in their columns. Then one William Pluright, grocer, advertised that in addition to groceries, patent medicines, smoked herrings by the box, raisins by the keg and Old Peach Brandy by the barrel, &c., &c., he also had for sale "Dupont & co's. gunpowder, superior to any imported."

Irénée du Pont had a long and rough road to travel before attaining even the small distinction of Grocer Pluright's paid notice. He was a "foreigner." He spoke little English. His masons and carpenters, gathered from Wilmington, Chester and Philadelphia and housed on the place, had difficulty in understanding him. Especially was this true when he had them build stone walls of thrice the customary thickness, and then place on them woodwork that was decidedly flimsy. No such powder-mills had ever been built in America. The idea which prompted this ingenious plan of extra-heavy walls of stone on three sides and light framework on the fourth side and roof was that if an explosion should occur, its chief force would be directed upward and toward the stream, away from other buildings.

Irénée puzzled his workmen with another innovation. Instead of putting up one large building to house the whole works, he erected a number of small buildings spaced well apart. The lack of this very simple precaution had spelled the abrupt end of many a gunpowder plant, among them America's first, that at Milton, Mass., in the year 1744.

The first winter was an open one. Sickness attacked the family. They missed the elder Du Pont's irrepressible optimism and, in this land of strangers, they longed for France. Spells of melancholy kept Irénée depressed for days. Details worried him, and, above all, the fact that he was always short of money.

Only a portion of the capital subscribed by the parent Du Pont Company had been paid. In New York, Victor had made heavy advances on account of the French expedition to Santo Domingo, for which he had been repaid only in part, and both Jerome Bonaparte and the Consul-General, Pichon, had drawn on him freely for funds. Two ships loaded with Victor's cargoes were lost during the winter. He, too, was fighting without capital, to pay notes which meant bankruptcy if unpaid. However, Peter Bauduy pledged his credit at the Wilmington banks and

Irénée kept building even as he shuddered at the rising debt that mortgaged his future.

By February, 1803, he was able to write his father:

"We have accomplished an astonishing amount of work since August, but I am dismayed when I think what is still before us. Within three months we have built a large house and barn of stone and a greater part of the refinery; we have repaired the water-course and the sawmill in which we prepare the wood for our framework, and a part of that used for the machines. This month we have still to build three mills and one or two other buildings; to dig a new race for one of the mills; to make the drying place, the magazine, the workmen's quarters. It is evident that we cannot make powder before autumn."

He worked incessantly, writing and planning into the night. "The activity of my work helps me and is good for me," he wrote in April, "in that it gives me less time to yield to the melancholy that never leaves me and that, I am afraid, affects my health."

With spring, he searched the woods for plants and trees new to him, wrote letters to French botanists and park authorities urging them to restock the depleted French forests from American seeds, and even proposed as a volunteer agent to supply the plants and seeds needed.

"My position in the midst of forests," he explained, "will make it an easy task and perhaps a valuable one. Some of the seeds will not be wasted and by these shipments and some successful work with plants and trees I may make for myself a position in France, and some day in the future secure a place in the Administration of Forestry."

Irénée du Pont was a powdermaker by circumstance. He was still homesick for France. In July, he rejoiced at the purchase of Louisiana. "The Federals are ashamed of their blustering of last year and are obliged to admit that Jefferson's methods are better, safer and cheaper than would have been a war with France and Spain, and a forcible invasion of the country by Kentucky Riflemen." But while he hoped for a kindlier feeling in America toward Frenchmen, he wondered if indeed it would ever come:

"In spite of the equality, the rights of liberty, and the excellent Government of this country, we foreigners are always in a position inferior to that of other citizens; we are not, as you say, among our equals; that is a truth that I have learned from daily contact with Americans. This suggestion of inferiority—this prejudice of which one often feels the influence—offsets in my mind many of the advantages of America and makes me believe that if we could be free from debt we would all be happier in France. . . ."

In midsummer, a full year after his arrival on the Brandywine, Irénée's refinery processed its first saltpeter. He wrote Jefferson of the event and shortly, in reply, received an order from Major-General Henry Dearborn, Secretary of War, to refine some saltpeter in the Army's possession. This was the first work done by the Du Pont Company for the United States Government.

Another autumn and winter passed, however, before the first black grains of finished powder came from the completed mills and were sent by boat to Victor in New York to be sold. At once Victor prepared a notice for the newspapers: †

E. I. DU PONT DE NEMOURS GUNPOWDER MANUFACTORY

Wilmington, Delaware

This new and extensive establishment is now in activity and any quantity of powder, equal if not superior to any manufactured in Europe, will be delivered at the shortest notice.
Samples to be seen at

V. DU PONT DE NEMOURS ET Cie

New York

At the suggestion of Du Pont *père,* the plant on the Brandywine was named Eleutherian Mills, which might be translated "Mills of Liberty." Irénée would have preferred the name "Lavoisier Mills," because they "would never have been started but for his kindness to me."

†There is no evidence, however, that this advertisement was ever published.

Growth—and War

NAPOLEON, crowned Emperor in 1804, kept Europe ablaze with war throughout the next decade. In America agriculture, shipping and every kind of business profited, either by the direct sale of goods abroad or by the lessened competition of European goods at home. The new Du Pont mills were launched on this rising tide of prosperity.

Output in 1804 amounted to 44,907 pounds of powder. In June of 1805, the business had so grown that outside sales agents were employed. In tests, the new powder surpassed that of the English and "left no stain on paper when burned."

American naval guns, belching defiance to Barbary pirates off Algiers in 1805, fired Du Pont powder for the first time in war—22,000 pounds of it. The result of this sale was such that Irénée could write to his father with pride:

"This powder was tested several times at Federal City* and was compared with powder sent by all other manufacturers, as well as with some lately received from England, and it proved so superior that old Mr. Dearborn, in spite of his unwillingness, sent us about 120,000 of powder to remake and a part of his saltpeter to refine, and he announced publicly on the Fourth of July before the officers—who were delighted with our powder

*Washington.

—that in the future we shall have all of the Government work. Aside from that I have in the last month made 40,000 of Army powder for the Spanish minister."

Trouble rode with prosperity, however. Victor went into bankruptcy, even as Talleyrand, too late, was personally pledging credit for him in Paris. Irénée begged his brother to join him in Delaware. Instead, Victor migrated to the Genesee Valley, to an outpost of the expanding civilization of New York State, where for the next three years he farmed and operated a general store. His family, augmented by a second son—Samuel Francis, born at "Goodstay" in 1803—spent the first year of Victor's absence at Eleutherian Mills. Here, in 1806, Victor's second daughter, Julie Sophie Angélique, was born. She was the first Du Pont born in Delaware, although all four of Victor's children were American-born.

True to promise, the chiefs of the French powder service sent Irénée a head workman, one Charles François Parent. Irénée was "dismayed" by his "gentlemanly manner." Parent objected to living "in the middle of the woods" and, in less than a year, plotted to break his contract and start a rival powder mill. To prevent this, Irénée had to lodge him in jail for three months, after which he gave him $1,000 worth of machinery and tools and sent him to New Orleans to start a powder mill of his own there. Even at the price Irénée considered getting rid of Parent "a great piece of luck."

A rift had formed and now rapidly widened between Irénée and Peter Bauduy. The latter had bought the shares in the company that had been allotted to McCall, which gave him a cash investment of $8,000 in the business. In addition, he had endorsed notes at Wilmington banks for $18,000. With most of his personal fortune thus involved, he became critical of Irénée's methods. He objected to Irénée's feeding the workmen and using them for work in his garden. Irénée's unending efforts to better his processes and powder irritated Bauduy, who believed the powder was good enough and should be more aggressively advertised and sold.

Du Pont did not want "the public attention on my establishment." He disliked "all that savors of self-praise and bragging—a method that seems to me more harmful than useful in any kind of business." Feeling that his powder should sell on its merits alone, he devoted himself to experiments that would reduce the danger of explosions from powder dust, in designing machines to speed manufacture and reduce its cost, and in seeking ways further to purify the product. He wanted "to make my mills not equal to any in Europe, but even better."

Irénée resented the authority Bauduy took upon himself and refused outright to have any name other than his own identified with the company. "I will not agree," he wrote Bauduy, "that an industry I have developed and to the success of which I have given ten years of my life shall be known by another name. If, as I hope, it earns a reputation greater than that of others and if it makes a name— that name should be mine."

Bauduy was worried by his notes and heavy investment.

Irénée was worried not alone by the notes but by the fresh memory of Victor's failure, which left him the one hope of his father's original company and the one support of the entire Du Pont family.

In France, his father had reported the powder company's successful start with eloquent—too eloquent—enthusiasm. The result was that the European stockholders were already demanding their share of the profits, which as yet existed only on paper. The mills had far exceeded the original estimates of cost, part of the capital was still unpaid, money for powder sold could not be collected short of six months, whereas the workmen expected their wages monthly and the banks were meticulous about their interest dates.

Tired of new construction, which meant new debts, Bauduy persuaded Irénée, against his better judgment, to burn charcoal and to dry powder under the same roof with only a stone partition between the two operations. The first evening after doing this, 800 pounds of powder were in the drying house. With the greatest care they extinguished the fire in the furnace of the adjoining charcoal room and went to supper.

Fifteen minutes after supper an explosion shook the place. Every window-pane in the house crashed. Dishes and glassware were broken, chairs overturned.

Irénée ran outside. The drying and charcoal house was a jumble of scattered stones from which shot flames. "Dalmas!" he shouted, sure that Sophie's brother was lost, but Dalmas answered in a moment. The building had blown up when he was only twenty paces from it.

Most of the night they fought the fire, every man of them knowing that in the graining house near-by were stored 12,000 pounds of powder that a single stray spark might send heavenward. Only the luck of a chill damp wind that blew away from the graining house probably saved the establishment and, no doubt, the company.

The business forged on. Sales in 1805 were $46,857.75 as against $15,116.75 the previous year. Income remained at about this figure in 1806 and 1807, although output and sales both increased, prices being lower. A new mill and other facilities were added to give a total output of 300,000 pounds of powder yearly. In February, 1808, Irénée wrote his father:

"I do not know yet whether we will have war or not, but at any rate I am ready, and our reputation is now so well established and our market so extended that even in time of peace we will have a larger demand than we can fill."

At the close of 1809, the book profits of the company for the period of its operation amounted to $43,613.68. Profits in 1810 exceeded $30,000, in 1811 approximated $45,000.

America's relations with England, meantime, had neared the breaking point. The likelihood of war had been a street-corner topic for three years. Influenced in part by this talk, in part by the example of Du Pont's success but most by the rapidly growing commercial demands for both hunting and blasting powder, a rash of new powder mills had spread over the country. The official census of 1810 listed more than 200 mills in sixteen states with an estimated annual output of 1,500,000 pounds. Yet demand still exceeded production.

Most of the mills were small and primitive; need had forced their existence. With no railroads, few canals, rivers that froze in winter and roads that often were impassable, the more isolated districts had to depend on near-by sources of supply. The Du Pont mills were listed as the country's largest; by coastal schooners and wagon teams the company shipped powder north to Boston, west to Pittsburg, and south to Charleston and Savannah. Irénée was confident that, with sufficient capital, he could have doubled his sales without making a pound of powder for war or export.

When war with England did come, in June of 1812, and the Government called for large supplies of military powder, its demands could be met only by sacrificing a substantial part of the rich commercial market and by increasing the Brandywine plant's capacity. Despite "old Mr. Dearborn's" Fourth of July speech in which he had promised Du Pont "all of the Government work," Irénée actually had been awarded only a small part of it.

From 1805 to 1809 inclusive, for example, the company's records showed less than $30,000 worth of Government orders as compared with award of $214,214.79 to other powder manufacturers. Politics and even kinships of high officials, as well as open prejudice against Du Pont's French origin, had almost blocked Irénée from Government contracts.

In other ways, too, the sudden war demand came inopportunely. In 1811, Irénée had loaned his credit to help his brother Victor, who had returned from the Genesee, to start a woolen mill on the opposite bank of Brandywine Creek. Bauduy and his son Ferdinand had also joined in the venture under the name Du Pont, Bauduy & Company. Almost at the same time, the parent Du Pont Company had failed in Paris with twelve shares of the powder company's stock as its only asset. The elder Du Pont had painted such a glowing picture of Irénée's success that the French stockholders, as one, demanded immediate payment of all profits credited on the powder company's books.

Vainly Irénée pleaded, in letters that often took six months to reach France, that his profits were represented not by cash but by raw materials, machinery, the long-term paper of customers and notes at the bank. The French stockholders threatened to liquidate the powder company. Thus, Irénée's credit was jeopardized when he needed it most. The Government had ordered 50,000 pounds of powder late in 1811. It ordered 200,000 pounds in 1812 and 500,000 pounds in 1813. On paper, orders of this size looked like strikes of gold, but, actually, they took Irénée onto the thin ice of possible ruin. To fill them, he was forced to risk every dollar he had in cash and to borrow heavily to extend the capacity of his overworked mills.

An adjoining property known as the Hagley Estate was bought to this end, at a cost of $47,000, covered chiefly by a mortgage.

The war, unpopular with many people, did not go well. In less than two months the entire Northwest Territory was in British hands. Before the end of 1812, United States troops failed in three attempts to invade Canada. The result was that manufacturers had great difficulty in collecting payments for supplies furnished the Government. In effect, they had to finance the war while the Treasury struggled to raise enough funds to avoid bankruptcy.

The situation grew worse as the war continued. In 1814, the English fleet blockaded the American coast as far north as Massachusetts, paralyzing all commerce and collection of revenue duties. Washington was captured, the White House burned. A Government loan failed. Most banks outside of New England suspended specie payments.

Even with the prospect of profits if the United States won, it took courage to support the War of 1812. This was especially true in Du Pont's case, although nowhere in his letters is there a trace of evidence that he hesitated one moment. On the contrary, he believed the war would result in vast good to the United States and should be prosecuted to the limit:

"Should the present war in its final result produce no other good but to secure the establishment of home manufacturers, so that we could manufacture our own produce for our own use, it would repay the nation ten-fold for every expense or loss the war may create. . . . The chance of purchasing goods at a low price

and selling the same to advantage in another country is very little
more than gambling on a large scale.

"Manufacturing, on the contrary, is a true creation of wealth.
It is taking cotton which costs 20 cents per pound and making it
worth several dollars. It is taking wool at $1 and selling it again at
$6 or $7 when it has received its metamorphose from our industry. Let us reap then the full advantage that this war may produce. Let us secure forever the establishment of American manufacture. . . . Almost every kind of goods can be manufactured
in this country as cheaply as in Europe. . . ."

Early in 1813, a British squadron under Commodore
Beresford sailed into Delaware Bay with the evident in-
tention of proceeding up the Delaware River to destroy
the Du Pont mills. When the people of Lewes refused their
demands for food and water, the British bombarded the
town; but so lustily did the Delawareans reply with
Du Pont powder rushed to them, that, after twenty-two
hours, the ships withdrew. The damage to Lewes was
slight, one account limiting it to $2,000 in injured build-
ings and "one chicken killed, one pig wounded, one leg
broken." However, the threat against the Du Pont powder
mills was real. Thereafter, troops were held in readiness
for their defense, and the workmen were organized into
military companies.

The mills' importance to the American forces was
greater than is indicated by the total of 750,000 pounds of
powder that can be definitely identified as military orders.
This powder was exclusively for the Army. The Navy
bought through its own agents, usually from local dealers
in the principal ports. Through such dealers, Du Pont sold

powder to ships of war and to privateers, contributing to such victories at sea as that of the *Constitution* over the *Guerrière,* the capture of the *Alert* by the *Essex,* of the *Frolic* by the *Wasp,* and the sinking of the *Peacock* by the *U.S.S. Hornet.* The American privateers captured more than 300 enemy ships during the first year and a half of the war.

Despite the loss of Washington, the tide of victory was America's in 1814. The British were beaten at the Battle of the Thames and again at Chippewa. The Northwest Territory was retaken. Over roads that were little more than forest trails, husky American teamsters had carted Du Pont's powder to Perry at Lake Erie.

Peace was signed at Ghent on Christmas Eve of 1814, but the news did not reach America until after Andrew Jackson had disastrously beaten the British at New Orleans in January. Du Pont's sales, both military and commercial, had fallen from $148,597.62 in 1812 to $107,-291.20 in 1813, and were still lower in 1814. Profits were actually less than the 1810–11 rate of strictly commercial earnings indicated they should have been had the war not occurred.

On the other hand, Irénée du Pont's foresightedness had saved the American Government a considerable sum. The Government bought and stored its own saltpeter. While Jefferson was still President it had bought on Du Pont's advice, against the eventuality of war, $50,000 worth of saltpeter from India at prices ranging from 16 cents to 20 cents a pound. This reserve was available when the war

shut off the Indian supply, and when inferior saltpeter
from the Mammoth Cave in Kentucky was selling at from
32 cents to 38 cents a pound.

Some indication of the difficulties that Irénée faced
during the war is given by a letter written June 20th, 1814,
in reply to urgings that he return to France. He wrote,
speaking of his personal affairs, not the company's:

"I do not see what I could do in France. I have spent my life
here building up a very difficult industry and the disappointments
I have had to bear have given me an habitual dullness and melancholy that would be very out of place in society. Besides, how
could I go? My position here is a very trying one; I have used
all the credit I could get to start Victor's establishment and to
buy sheep; and neither of these enterprises has been nearly as successful as the powder. I owe more than sixty thousand dollars,
chiefly in notes at the banks, so that my debts amount to far more
than my profits from the powder. I am forced to stay here; the
signatures that must be renewed every sixty days put me in
exactly the situation of a prisoner on parole who must show himself to the police every month."

Bauduy wrote the next day, fearing ruin: "We now
have 140,000 dollars invested in buildings, water powers
and land on the Brandywine—a piece of madness that I
was unable to prevent."

The company's books as of June 30th, 1814, showed
$125,000 tied up in raw materials and finished powder,
debts in excess of $100,000, and $10,193.08 cash in bank.
Receivables were slow in coming in. The United States
still owed more than $7,600 to the Du Pont Company in
March, 1815, almost fifteen months after the Peace of
Ghent.

Road's End

Both Irénée du Pont and Peter Bauduy were devoted fathers. When, therefore, Ferdinand Bauduy married Victorine du Pont in November of 1813, the union seemed to augur a lasting partnership between their parents despite the conflicts over business policy. But only a few weeks later young Bauduy was stricken with pneumonia and died.

Bauduy had objected strenuously to Du Pont's bold expansion of the mills to supply powder for the War of 1812. He wrote to the French stockholders, already angry because they had been paid no profits, charging Irénée with waste and incompetence. When these letters came to light, Du Pont forthwith made Bauduy a cash offer for his stock and gave public notice that the partnership was dissolved.

Repercussions of the break reached Europe. Banker Jacques Bidermann, one of the chief stockholders, immediately sent his son Antoine to America to make a personal investigation. Young Bidermann was only twenty-four, but he had an inherent shrewdness and a good knowledge of accounting. Soon he was at the Du Pont office delving into the books.

Du Pont *père* was alarmed. Napoleon's abdication as Emperor in April, 1814, and the return of the Bourbons to the French throne, had led to his appointment as Secretary of the Provisional Government, then as Chevalier of the Legion of Honor and Counsellor of State. At the moment, he was too occupied to leave France, but he was impatient to return to "the Brandywine." He was America-bound sooner than he expected. Napoleon escaped from Elba and on March 1st, 1815, landed at Cannes, en route to Paris and the One Hundred Days that were to end at Waterloo. In May, the elder Du Pont, now past seventy-five, bothered by gout but with the zest of old, was in Delaware again to the delight of his sons and grandchildren.

In the interval, young Bidermann had proved to his own satisfaction that Irénée had treated the stockholders not only fairly but generously. Instead of the $22,000 pledged him by the original company, Irénée had received from it only about $16,500. The profits that he had been withholding he had credited to the stockholders' unpaid balances, with compound interest, for reinvestment in the business. As a result, the stockholders' original investments had more than quadrupled in value. Bauduy's original $8,000, young Bidermann figured, had netted him in dividends and commissions more than $100,000 in twelve years.

The moment these facts were clear, Bidermann joined in ousting Bauduy from the business. Du Pont *père* arrived to find the young Frenchman occupying Bauduy's old desk and serving in his place as sales manager. Bauduy had drawn about $40,000 since the start of the business, according to the books, which left about $60,000 due him in the final settlement. Of this $37,714.28 represented his original investment of $8,000 in stock and its accumulated earnings to the fifteenth of February, 1815. That is to say, each $2,000 share of stock in the company was now worth $9,428.57.

The elder Du Pont was delighted. He was amazed at what Irénée had accomplished in thirteen years and at the progress of the country since 1800. Du Pont powder sales in 1815 exceeded by 75,000 pounds those of the biggest year of the war. Such was the nation's demand for powder of all kinds that "sweepings"* sold at $8 and $9 a keg. Young Bidermann wrote from New England as early in the year as March that "there is no powder . . . the purchasers do not ask for a good quality—only something that looks like powder."

Talk was no longer of Europe and her wars but of the American "West." Farmers needed powder to clear their fields of stumps and boulders and to protect stock from predatory animals. The fur industry took large quantities. But real money, cash, was lacking—the "West," too, was being built "of paper." Credit often had to be extended a year or longer. Banks were snowed under by notes.

Prices on most manufactured goods were ruinously low. In a great post-war drive to overcome the formidable American competition that the war embargo and blockade had fostered, British manufacturers dumped shipload after

---

*A technical name for the fine powder dust swept from the floors of the mills and generally returned to be reworked with the new ingredients entering the processing.

shipload of goods into the United States at prices less than production costs. Europe's armies, disbanded, no longer bought American wheat. The country headed toward the inevitable panic that follows in the wake of war.

Irénée du Pont fought, now, the most trying battle of his years in America. He fought for his company's, his family's and his own economic life. Paying off Bauduy had taxed the company's resources. Now Bauduy built a rival powder mill a few miles away and raided the Du Pont working forces to man it. He copied Irénée's machinery and methods and coaxed away some of his sales agents. He denied the adequacy of the settlement made with him and instituted a lawsuit that was to drag through the courts for eight years before being decided against him.

On June 8th, 1815, an explosion killed nine men and caused $20,000 in property damages to the Du Pont mills. It was the first fatal accident. Two years later, through the carelessness of a workman, fire started in the charcoal house and it burned to the ground. This time the victim was Pierre Samuel du Pont de Nemours.

The old gentleman was retiring for the night when the alarm was shouted. Pushing aside all who would restrain him, he joined the bucket brigade. He was soaked, singed and blackened, but worked as zealously as the youngest man in the line. Next day, hoarse, painracked and exhausted, Du Pont de Nemours kept to his bed. He made light of his condition, but on August 7th, 1817, he passed on to his last and greatest adventure.

Jefferson wrote that "A beloved friend, a patriot, and

an honest man" had died. Today, a simple stone marks his final resting place in the family burial ground near Montchanin, Delaware.

Irénée du Pont was now forty-six, one year older than his father had been that day at Bois-des-Fossés when he had girded on the boy of fourteen the sword of manhood. Although the younger son, Irénée was now the head of the family. Gray flecked his hair. The wistful, shy eyes of boyhood had become the deep, sad eyes of a man who throughout life had known little rest. His burden was to be no lighter.

The facile pen of Du Pont *père,* while he lived, had held off the impatient French stockholders. His death removed their last restraint. Loudest and most persistent of all was Irénée's stepsister, Madame de Pusy. He offered to pay $12,000 for her $2,900 investment in the business, in addition to upwards of $10,000 she had already collected over the years from profits, but she joined Bauduy in his court action to dissolve the company. Just ahead of Irénée, too, waited another disaster.

On the morning of March 19th, 1818, he was in Philadelphia. In one of the five mills of the old, or Upper Yard of the powder works, several men were at work—never were more than a few permitted to assemble in any one place because of the constant hazard. Suddenly the earth was shaken by an explosion that hurled the mill skyward in a blast of smoke streaked with flame. Two other explosions followed. The heavens, blotted out by smoke, rained fragments of stone and firebrands over several acres. From everywhere into the open rushed white-faced men. In their panic they converged. The next instant the earth was wiped bare as the fourth and fifth mills of the yard exploded. Thirty-six were dead, four injured mortally.

Irénée du Pont galloped an exhausted horse home to the wreck of half that he had built. His wife had received a blow in the side, from the effects of which she never recovered. Dalmas had a broken arm and other injuries. Victor was whole, but a man at his side had been killed instantly. The houses of the workmen were crushed, although they had stood a half-mile from the exploding mills. More than 85,000 pounds of powder had blown within four minutes. Windows had been broken in Wilmington. The shocks had been felt even in Lancaster, Pennsylvania.

Irénée found many of his best workmen among the dead or permanently maimed. Every precaution he could contrive had been taken to prevent just this thing, but little, he reflected sadly, could be done for these men now. However, in a day when laws to protect labor were notable for their absence, he set a precedent for all who might follow him. The widows and the maimed were pensioned. They were given houses in which to reside as long as they lived. Provisions were made for adequate medical care of the injured, and for the care and education of any children left parentless.†

† "By this calamitous event," wrote Benjamin Ferris in his *History of the Original Settlements on the Delaware,* published in 1846, "many

Then Irénée began to rebuild, even though depression had settled upon the nation, with banks failing and agents and customers bankrupt. For a year, not a pound of powder could be produced in the Upper Yards. It was almost impossible to get new workmen. Many old workmen quit "the Brandywine."

At the end of 1819, Du Pont estimated his losses over two years from bankruptcies, explosions and other causes at $140,000, and from deterioration of values of real estate, because of the general financial distress, at $50,000 more. His debts seemed insurmountable. By every rule of business, excepting one, the powder company was doomed. That rule was the man himself. Banks were still ready to lend money on the security of his signature.

The notes built new and better mills. The disaster, too, served to temper the demands of the impatient French stockholders. They agreed to accept long-term notes in lieu of cash due them. Throughout the balance of his life, however, Irénée was to plod the treadmill put under him

---

were left widows and helpless orphans. In this case the benevolent disposition of E. I. Dupont had ample room for exercise. He pensioned every widow and provided an asylum for every orphan at his own expense; furnishing them with clothing and the means of education."

Company correspondence suggests that victims of accidents may have been compensated by Du Pont as early as 1815. On June 16th of that year Bidermann wrote Du Pont from Pittsburg: "Sandrans' letter was to tell me of the terrible accident to the stamp mill. I agree with you— the loss in money and time does not amount to much, but I am very unhappy about the men; it is horrible to think that they died in our service. I think the company ought to give a pension to each of their families; it is our duty to do all that we can to atone for their loss. If you agree with me, I will be glad to subscribe on my own account and my father's to whatever you think best to do."

by these commitments and by his voluntary assumption of the debts of his father's defunct company.

During the Christmas season in 1822, Cæsar A. Rodney, United States Senator from Delaware, informed Irénée du Pont of the Senate's confirmation of his nomination by President Monroe as a Director of the Bank of the United States. Mr. du Pont accepted the appointment, and later his reappointment by John Quincy Adams both as a responsibility and an honor. The foreigner of 1802 and the homesick man of subsequent years who had wished hopelessly for a position in the French forestry service had disappeared. In his place was an American, proud of his citizenship, intensely interested in the land of his adoption, consulted by those in authority on legislation to help manufacturing and farming.

When La Fayette made his triumphal tour of the United States, in 1824–25, one of the homes he chose to honor as guest was that of Irénée and Sophie du Pont. The home was the same modest stone house that they had built in 1802–3, along with the first powder mill; it held almost the same furniture; and only thirty paces away across the front yard was the same "home office" where, as usual, Irénée worked into the night. Like the youth La Fayette had known in France, the man still sought security.

"It is cruel," he wrote Sophie on one of the interminable night's absences in Philadelphia, "to ride sixty miles every five or six days to meet one's notes, and so to waste one's time and one's life. God grant that some day I may get to the end of it."

He never quite did.

Irénée saw beyond gunpowder. In addition to helping establish Victor in the woolen business, he aided in setting up a cotton mill and later a tannery. To none of those, however, was he able to give of his own time. One after another they failed. At every opportunity he urged a closer alliance between manufacturing and agriculture, so that the factory might become an ever larger consumer of the farmer's products. His uninterrupted work with merino sheep did much to better the quality of American-grown wool.

The chemical base of the powder company was broadened under his direction by the addition of a considerable line of chemicals. Two of these chemicals, creosote and pyroligneous acid, were derived as by-products of the distillation of willow wood for charcoal. By immersing scrap iron in pyroligneous acid yet another chemical was obtained, known as black iron or iron liquor, which was used in the dyeing of leather and calico. These represented one of the first efforts in America to utilize the so-called waste products of industry, which on occasions have been found to be of greater value than the parent product. Refined saltpeter and charcoal contributed to this chemical line.

Before the War of 1812, good imported powder sold in the United States at 40 cents a pound. Its price in 1827 was from 26 to 30 cents, but American powder of equal quality was priced from 16 to 20 cents. More than any other one man, Irénée du Pont had brought about that change. The Du Pont mills in 1827 employed 140 men, had a capacity of 800,000 pounds of powder annually, and ranked with the best in the world.

That year, Victor du Pont was stricken ill on the street in Philadelphia, dying an hour or two later in the United States Hotel. Down through the years Irénée and he had, indeed, been "doubly brothers," although opposites in tastes, temperaments and talents. In business, Victor had failed. In diplomacy he might have ascended the heights. He died in his sixtieth year, a member of the Senate of his State, a man whose friends were everywhere.

Then Sophie was taken ill. Irénée's usually full and courteous letters to his agents became abrupt, at times only a few scrawled lines.

August 3, 1828

"I am not able to attend to business affairs as I should. My only thought is for my wife, who is painfully and dangerously ill and has been so for several weeks. I am unable to work or sleep."

August 5, 1828

"Mrs. du Pont is so ill today that I hardly know what I am writing."

Hours he sat by her bedside. It was he who administered the medicines prescribed by the doctors. She seemed to improve as September brought relief from the summer heat, but he was afraid to go beyond call of her voice. His

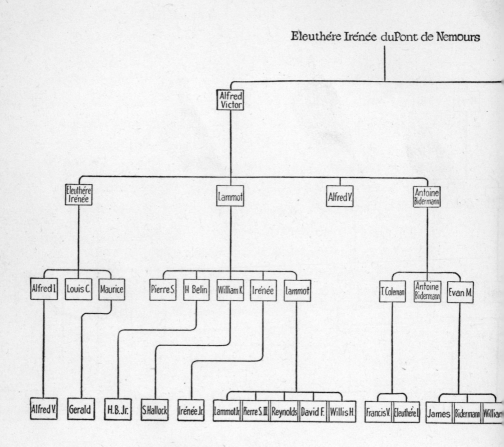

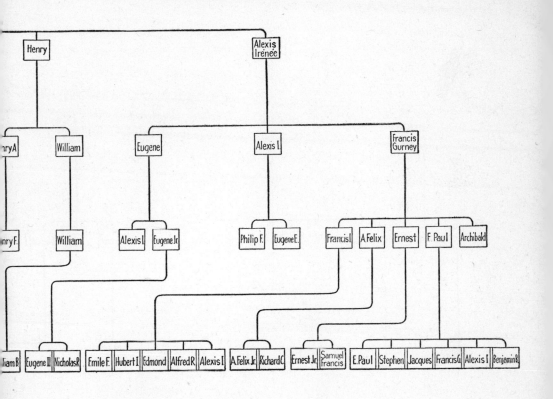

letters grew fewer, finally ceased, and Bidermann handled all business. On November 27th, 1828, Sophie Dalmas du Pont followed Du Pont *père* and Victor.

After that Irénée found solace only in work. Work meant for his children that security he himself had yet to find. The widowed Victorine took her mother's place as mistress of the household. Evelina had married Antoine Bidermann. Alfred, Irénée's eldest son, now thirty, was an assistant in the mills, intensely interested in chemistry. Four other children—Eleuthera, Sophie Madeleine, Henry and Alexis Irénée—had been born at Eleutherian Mills. In a short time Henry, sixteen, was to enter West Point, and Alexis, twelve, a boys' school in New Haven, Connecticut.

By nature a quiet man, Irénée grew quieter, grayer. Occasionally, in writing letters, he had lapses of mind from which he would recover with a tired smile to find he was writing in French instead of the now customary English. However, he saw his debts yearly grow less, the business greater. He was proud of Alfred, whose chemistry was improving his best powder.

In 1832, at the request of the Treasury, Irénée reported to Congress an estimated "present value of land, buildings, water powers and machinery" of $80,000, a working capital of $100,000 and an annual output of 850,000 pounds of powder, which he believed to be about one-seventh of the total American output. He reported Du Pont powder was sold exclusively in the home market, although the industry exported about 1,200,000 pounds a year. The grand total of all powder produced by the Du Pont mills in the thirty years since their founding he placed at 13,400,000 pounds. The bulk of this had been used for hunting, which demanded a powder superior to blasting powder. Hours of work at Eleutherian Mills, as usual then in American industry, were twelve hours a day in summer and nine hours a day in winter—eleven hours a day average for the year—daylight to dark.

Slowly but certainly throughout these years a breach had opened and widened between North and South. The "tariff of abominations," passed in 1828, had caused the agricultural South to seethe with indignation. This, by 1830, had boiled into open threats to nullify the Constitution. Jackson's election to the presidency brought a lull, but in 1832, a new tariff act was passed that was again obnoxious to the South. In November, a State convention called by South Carolina declared the new act would be ignored by that State after February 1st, 1833; the State legislature authorized funds for the purchase of arms and the raising of a military force to resist Federal enforcement of the law. Jackson answered by moving troops and ships into the vicinity of Charleston.

At this juncture, an order for a large quantity of military powder reached Du Pont for shipment to the irate South Carolinians by way of Philadelphia. Irénée du Pont's answer to his agent went back by the first post on January 12th, 1833:

"We have duly received your favor of yesterday transmitting an order from Mr. C. R. Holmes, of Charleston, South Carolina, for 120,000 pounds of cannon, musket and rifle powder, to be paid cash and shipped by the 24th inst. at Philadelphia on board a vessel for Charleston. The destination of this powder being obvious, we think it right to decline furnishing any part of the above order. When our friends in the South will want sporting powder for peaceful purposes we will be happy to serve them."

Thus precedent was set for those who would follow Irénée du Pont. It was knowingly to sell no powder for use against the Government of the United States.

Another year rolled around. "I am not sick," Irénée wrote in August, "but weakened by the heat and tired from too much exertion." Irénée's work was almost done. He checked over the last of his debts. Only a few were left to be paid. Share by share, through the years, the stock held in France had been taken up until most of it was now locked away in the company's safe. At times he dreamed of going back to France for the final settlement when the last claim there would be canceled and his parole ended. That, too, was left to be done.

In Philadelphia, toward the close of an October day, Irénée stumbled as he made his way back to the United States Hotel. A block further on he fell and was unable to rise. Passers-by carried him to his hotel room. He died before dawn, October 31st, 1834, of a ruptured heart,‡ in the same manner and in the same house that Victor had died.

"No event within our recollection," wrote the editor of

‡This account of his death is from the *Delaware State Journal*. Other accounts say that he died of Asiatic cholera then prevalent in Philadelphia.

the *Delaware State Journal,* "has spread a deeper gloom over this community than the sudden death of this excellent man. . . . On Friday morning it was rumored through the town that an express had passed through in the night to his family, bringing intelligence of his sudden and extreme illness in Philadelphia. The anxiety of our citizens until the morning steamboat arrived from Philadelphia was intense—crowds assembled on the wharf to learn the event; and when the fatal intelligence was announced of the loss of one so esteemed and so loved, each seemed to feel it as a blow inflicted upon himself. . . . We have lost a friend whom we loved and venerated, this community a benefactor—our State its most useful and valuable citizen."

But the Eleutherian Mills, in 1834, produced more than 1,000,000 pounds of black powder. The institution lived!

VIII

Change

THIRD of E. I. du Pont's children and the eldest of his three sons, Alfred Victor du Pont was thirty-six years old upon the death of his father. That event came to him as a shock and as the forerunner of a duty from which he shrank by temperament. Alfred had no desire to assume the company's leadership. He preferred a rôle in the background.

Literally, he had grown up in the mills. The "yards" had been his playground, workmen's sons his playmates. By night he had slept within a few hundred yards of the works—as a farm boy sleeps close to the paternal barn—with the smell of sulphur, saltpeter and charcoal in his nostrils. He had heard the dread alarm of fire, the crash of exploding mills. He had seen grief in the raw, and death.

At twelve years of age, Alfred knew every piece of machinery in the mills, every step in the processing of powder and its ingredients, every workman by his name. At sixteen, he was sketching machinery and experimenting with chemicals, a youth with an insatiable curiosity about better, and, above all, safer ways to make powder. At twenty, he was a workman in the plant, fortified by two years' training in chemistry gained at Dickinson College under the instruction of Thomas Cooper, an Englishman, the most eminent chemist in America at that time. In one of the buildings he set up a small laboratory where, when the mills were silent, he reveled in the hours of work alone, experimenting, testing, scheming. His great aim was to replace guesswork with precision, to eliminate the ceaseless hazard, to inject science in what had been for centuries largely an empiric art. Never robust, often in poor health, he recoiled from the constant meetings with strangers, the interminable letter writing and the enforced absences from the mills that general management would impose, now that his father was gone.

At this stage, therefore, Alfred was grateful for the presence in the company of his brother-in-law, Antoine Bidermann, now a veteran of twenty years in the sales and business ends of powder-making. Alfred insisted that Bidermann, temporarily at least, become the company's head. Consequently, from November 1st, 1834, until April 1st, 1837, the acting chief of the Du Pont firm was a man not a Du Pont by birth or name.

The arrangement proved a happy one. It gave Alfred time to familiarize himself with administrative work and to train his younger brothers in the manufactory. The second brother, Henry, now twenty-two, had been graduated from West Point and had seen service with the Army in the Creek Indian country. Only a few months before his father's death, at his request, he had resigned his commission to become a powderman. Alexis, eighteen, left school also to work in the mills. Into his brothers, with infinite patience, Alfred drilled his own and their father's creed.

So ably did Bidermann manage that in the spring of 1837 he went to France and paid the last dollar owed the original stockholders, and in the final settlement of accounts he included his own interests in full. Then he retired, leaving Alfred, Henry, Alexis, and their four sisters, Victorine, Evelina, Eleuthera and Sophie Madeleine, the sole owners of the mills. The brothers and sisters met in the company's office and, at their father's old desk, Alfred drew up a partnership agreement that was to endure for sixty-two years.

The company name was retained under it. However, the business was to have no officers; the three brothers, as its active managers, were to assume no titles. Profits and losses were to be shared alike. Everything belonging to the firm was to be jointly owned and apportioned according to each of the partner's needs, including their homes and the farmlands, the farm products, the horses and carriages. No salaries as such were to be paid, but each partner was to draw as needed from the firm's cash, the withdrawals to be deducted later from his or her share of the profits. As the eldest, Alfred was to speak and act for all seven, sign all letters, draw all checks. When there were differences of opinion, his decisions were to be final. Henry and Alexis were to attend to the operation of the mills.

Until almost the turn of the century, in accord with this agreement, Du Pont women were to participate in all important business meetings on equal terms with Du Pont men. More than one sound action was to be motivated by feminine counsel.

In the decade that followed, prosperity and depression mingled in the country. Completion of the Erie Canal in 1825 had reduced freight rates between Buffalo and New York from $100 a ton to less than $8 a ton. Four years later, the first practical locomotive, Stephenson's *Rocket,* had been successfully demonstrated in England. A new era opened in transportation as the nation awoke to the wealth to be tapped through faster and cheaper ways of shipping goods. In 1840, more than 2,000 miles of American railroads were in operation, canals were cut, steamboats plied back and forth on most of the great rivers. National highways penetrated the former wilderness. Southwestward from St. Louis covered wagons rattled and creaked toward the sunset and the gold and silver that awaited at the end of the Santa Fe Trail. Shortly the crack of whip and rifle were to make the Oregon Trail another pulsing artery of empire for years to come.

The Machine Age had dawned. Manning's mowing machine, Hussey's reaper, Moore's harvester and the iron plow were facts in 1837, when Alfred du Pont took up the reins of company management. Morse's magnetic telegraph was five years old, Faraday's dynamo a year older. Now came Goodyear's discovery of the art of vulcanizing rubber, Thompson's pneumatic tire, the turret lathe, the sewing machine, Hoe's rotary printing-press, penny newspapers. Home industry was passing to the factory, cities were growing at thrice the pace of the country as a whole, and as the long lines of covered wagons rumbled westward, Europe poured its hundreds of thousands into the industrial East in a fast-rising tide of immigration.

This varied activity of the country spelled business for the Du Pont mills. The population flowing over the Mississippi in wave on wave spread into every part of the Great West—and looked to gunpowder for meat and security. Railway and road construction, the growing use of coal for fuel and of iron for machinery made blasting powder indispensable. Yet, paradoxically, Alfred and his brothers worked under skies not dark but black. Speculation, boom and easy credit had stretched the banking resources of America to the limit during the early Thirties. One month and ten days after Alfred took charge of the Du Pont business, the banks of New York City suspended specie payment, soon to be followed by those throughout the country, and a great economic crisis confronted all business in the United States.

Almost one-third of the nation's workers had no jobs. Poor-houses were filled, the streets swarmed with people begging for bread. Hundreds died of cold and starvation. Except for a brief respite in 1839, conditions of the sort continued for seven years. Debts were repudiated. Throughout the country, 33,000 business houses went bankrupt with losses estimated at $440,000,000, a tremendous total for the time. By 1841, one-third of the banks in Ohio had failed and, it was said, practically every piece of property in Alabama had changed hands.

Most of Alfred's time during this period was taken up

with the grind of finance, work that he detested. As of old, business was to be had, but it was "paper" business. The unemployed swelled the western migration and to them gunpowder was a first need. Men who could not work to earn food took guns and hunted it in the woods. When weaker businesses failed, the remnants of their customers drifted inevitably to the stronger survivors. Among the survivors Alfred encountered a competition that showed no quarter. Blasting powder often sold at less than the cost of delivering it.

The wooden kegs in which powder was shipped were bought in Boston and Philadelphia. They were of all sizes and qualities, causing confusion in sales and frequent losses resulting from breakage or dampening of the powder. Alfred built his own "Cooperage Shop," designed a satisfactory standard keg and, thereafter, manufactured all kegs on the premises.

Coastwise schooners still carried the bulk of the powder destined for New York and New England. This powder had to be hauled by wagon-team to Marcus Hook, on the Delaware River eight miles above Wilmington, where it was loaded on small boats and rowed to the ships. Alfred improved this condition by building a magazine at Edge Moor, a shorter haul by five miles, and by erecting a pier along which the largest schooners could dock.

Specially designed covered wagons, drawn by four horses or six mules, according to the length of the journey, were added to the company's equipment to supplement the deficient facilities of railroads and waterways. Between Philadelphia and Pittsburg, for example, the only stretch of railway in the Forties was an incline road operated by cable over a part of the Allegheny Mountains. A single-track line between Hartford and Bridgeport was the only railroad linking Boston and New York. The Du Pont powder wagons were slow, but they went wherever horses and mules could get foothold to draw the load. And they reduced shipping costs.

May of 1846 brought the Mexican War. As far back as 1802, Du Pont de Nemours had foreseen the eventuality of this conflict and had warned Jefferson to "uproot from your nation" the temptation to despoil a weaker neighbor. Just as vigorously his grandsons opposed the war, but, once the United States was committed, they supported it.

During the war, the total amount of Du Pont powder furnished the United States was about one million pounds, or less than one-fifth the mills' estimated output for the period. The powder was supplied at a tragic cost. The Army's powder orders were urgent, necessitating the hiring of new, inexperienced men, work by lamplight, and the sudden augmenting of the military division of manufacture. In the rush, somebody was careless. A spark ignited powder in a press. Eighteen dead were found in the debris of the wrecked mills.

Alfred du Pont was a broken man after that. He watched Henry and Alexis rebuild. He saw the totals in his sales book grow—that for 1849 exceeded by 400,000 pounds of powder the record set by 1848. Day and night now, to meet the mounting demands of the prosperity that came with peace, the mills turned out the shiny black grains that went by wagon, boat, rail and ship to the farthest outposts of the nation. Alfred retired in 1850, a semi-invalid. He died in 1856.

With his going, more than an individual passed from the business on "the Brandywine." With him went the last influence of Europe and the eighteenth century. For, although Alfred had lived all but two of his fifty-eight years in the United States, the greater part of his life had been spent in a home atmosphere that was still of the Old World. He was French-American. His brothers and his sons were wholly New World by birth, environment and schooling. English was the first language they had learned, America's the first history they had read. They were Americans in views, temperaments, characteristics, prides, ambitions. They cherished the loyalty and love of family that was a part of their heritage, a trust passed on by father to son and to grandson, but with them a new regime came to the Du Pont powder mills. Now Henry, the West Pointer, took command.

"The General"

BY HIS own account, and he was a militarily exact man, Henry du Pont was of "five feet ten inches scant height," but he looked more than six feet in the glistening black "stovepipe" hat he wore for every occasion, except when wintry blasts forced him into woolen cap, ear warmers and muffler. That hat and the habit of wearing it at all times was to save a glazing mill one night in 1884. A shaft became overheated. It was showering sparks when Mr. du Pont discovered the danger. Running to the near-by creek, he filled his hat with water and doused the shaft.

Vigorous and robust at thirty-eight, shrewd of eye, with a fighting cleft chin around which, as he grew older, he cultivated a cropped beard, this American-born son of Eleuthère Irénée took up his brother's relinquished gavel in 1850 and wielded it with a sound that the whole powder industry was to hear during most of the remaining half of the century. He gloried in responsibility. Competition was like wine to his blood.

Every person for miles around knew "Mister Henry" or "the General," a title conferred by virtue of his appointment in 1846 as Adjutant-General of Delaware and later, in 1861, as Major-General commanding Delaware's volunteers. The General had three prime interests: the powder business, politics, and farming. The first he indulged as a heritage, the second as a patriotic duty, and the third as a passion. His cattle, horses and broad fields of wheat, his greyhounds that accompanied him even to work, his buggy on which rain or shine the storm curtains were up, were as famous and familiar as his beard, greatcoat and his stovepipe hat. His Spartan personal tastes and penchant for economy were to become bywords in Delaware. Weeds were anathema to him; in his buggy he always carried a trowel and woe to the daisy that took root in his trim fields.

Long after steel pens were in common use the General clung to his goose quill, with which, a greyhound at his feet, he wrote an average of 6,000 letters a year, mostly at night. Though others might work by coal-oil lamp or gaslight, three candles—no more, no less—filled the General's illumination needs. His invariable ritual was to light first the tall new candle on the left, then to light the used one in the center which had been his working companion the night before, and finally the short one on the right, which on the morrow would do thrifty duty in the old square lantern he kept on the window ledge ready for his return to the house.*

The typewriter had been invented ten years before he would tolerate the "noisy contraption." Throughout the almost forty years that he was to dominate the company and his family more completely than any other one man

*"Du Pont Romance," by George H. Kerr, privately printed, 1938.

before or since, the floor of his private office—just thirty paces from his home—was as bare as that of an Army barrack; the battered desk and wooden chairs were the same that his father had brought from France.

Not once in the four decades did he allow his desk to be moved, or the position of his chair changed, until the floor planks were worn hollow. A Wilmington banker once remarked that the General "carpeted his office floor in greyhounds" and added: "Henry makes damned well certain that the greyhounds reproduce themselves." It was a story the General told with gusto. He could laugh the loudest over stories told at his expense.

He knew people not casually but intimately. Valuable papers of workmen and neighbors—deeds, contracts, wills, marriage certificates, bonds—were kept in a special compartment in his office safe. He knew who was sick, expecting a baby, planning to marry, or who got drunk on Saturday night. He long served his home district as inspector of elections and challenger at the polls. Henry Clay was his political idol, a statesman, in his opinion, second only to Washington, and he smoked only "Henry Clay" brand cigars. The one unpardonable sin of citizenship, he felt, was a failure to vote the Republican ticket.†

General du Pont's first act upon taking up the company's management was to admit into the partnership Alfred's oldest son, the second Eleuthère Irénée, who at twenty-one was still serving an apprenticeship in the mills

†Then Whig.

under the eye of his Uncle Alexis. This formality over, the General began a task he had long awaited, a study of the company's books, which always had been open only to the head of the partnership. He suspected the firm's financial position was bad, but he wanted to know the exact facts.

Hours later, his desk littered with papers, his cigar long dead and forgotten, he closed the books, frowning thoughtfully. The powder mills were running night and day, their output exceeded 10,000 pounds daily, the quality of the powder was unrivaled, the manufacture efficient, and the company was the largest producer in the country. Yet the partnership owed more than $500,000!

Carefully the General restored the books and papers to the safe, each to its precise place. He lighted a fresh "Henry Clay" cigar. Then, with hat tilted over his shrewd eyes, his feet propped up, and alone with his greyhounds, he planned his campaign to avoid bankruptcy.

The great new market of powder was in mining, especially in Pennsylvania's hard-coal region. A cheap, low-grade blasting powder was sufficient for this work. Such powder, he knew, was being produced readily in competitive mills that had lower plant investments and less skilled workmen than the Du Ponts. Many of these mills, too, were nearer the mines and so had lower delivery costs. The General decided that, while the chief Du Pont emphasis for forty-eight years had been on quality, that now was secondary to price in the principal powder market, coal mining.

This much clear, the General wasted no time on further bookkeeping post-mortems. He declared Du Pont powder was not only good enough but as good as present knowledge and science could make it, that henceforth first emphasis would be on price, sales, and service. Alexis, the production chief, was told to search the mills for economies. If they expected profits, the General said grimly, they'd have to salvage them out of costs.

On the credit side he had found uncollected accounts on the books that dated back to his father's time. Politely but pointedly he called for payment. The company's agents, long accustomed to easy methods, were jolted into new life as the senior Du Pont hurried from city to city. New agents were recruited, men of energy who were good salesmen. They were stationed at every rising center in the new Southwest and West.

Discovery of gold in 1848 had started a stampede of the adventurous to California. They carried guns as commonly as they wore boots. Moreover, as one-man placer mining gave way to larger operations directed by engineers, blasting powder superior to that used for Pennsylvania coal was needed in California's gold mines regardless of its cost. Shortly, ships began to load at Du Pont's Delaware River pier that were of a class different from the coastal schooners that carried powder to New England. Long, low, with rakish masts and billowing sails, they headed down the South American Coast, around the Horn and up the Pacific with straining canvas and foaming wake. These clipper ships made San Francisco in about three months. There, too, the General's long arm had placed an alert Du Pont sales agent, his territory the whole Pacific Coast.

The General suspected that most of his competitors were no better off financially than his own firm. He went to them boldly with a proposal that the price war be ended. He negotiated what he called an "economic arrangement." It led to a return to prices more in accord with each producer's manufacturing and delivery costs.

Young Irénée made a notable contribution to economy. Why not, he asked, ship powder in metallic kegs instead of those of wood? The metal kegs would withstand more abuse, be weather-tight, safer, cheaper. He designed such a keg, patented it and the ancient wooden powder kegs passed into history—not, however, without a tragic finale.

A spring day in 1854, three big Du Pont powder-wagons came rumbling along one of Wilmington's main streets on their way to the Delaware River pier. They contained 450 kegs of powder. Something happened, what or how nobody living ever knew, for in one blast kegs, wagons, horses, drivers were blown to bits. Wilmington shook from end to end. Figuratively so did the nation. Towns and cities prohibited the hauling of powder over their streets. It was a wise restriction, even if it increased the difficulties of powder transport.

The Crimean War started about this time. Russia faced the combined forces of Great Britain, France, Turkey and Sardinia. More than 325,000 men were soon locked in a conflict, which was to consume gunpowder faster than the mills of Europe could produce it. Until now, the Brandywine mills had supplied no powder for Europe's wars, but this was an opportunity to bolster the firm's finances that General du Pont could not overlook. Upon the State Department's assurance that the United States was in no way concerned, Du Pont powder began to move to the Crimea by the shipload.

The bulk of this powder went to the English and French forces, the Du Pont mills supplying probably one-half of their heavy purchases in America. The English fleet patrolled the Atlantic in such efficient manner that Russia was practically blocked from the American market, despite the premium offers of its agents. The war business wiped the Du Pont books clean of debt.

To at least one young Du Pont, filling war orders was but an interlude in more important work, aimed, no less, at effecting revolution in the ancient powder industry. This member of the family was Lammot, the second son of Alfred Victor. Born April 13, 1831, and named for the family of his mother, the former Margaretta Elizabeth Lammot,‡ young Lammot had been graduated from the University of Pennsylvania as a chemist at the age of eighteen. Six feet two, lanky, big-boned, he had his mother's dark brown hair and a chin, mouth and cool gray eyes as determined as those of his Uncle Henry.

By the time Lammot was twenty-one, his practical experience in the mills plus his technical schooling and natural ability had given him a knowledge of powder-making to which both of his uncles and his elder brother

‡An Americanization of La Motte.

were quite ready to defer whenever any question arose. General Henry's word was the last on policy and sales, but Lammot's soon became the last on all problems involving chemistry, although he was yet to attain partnership in the firm. So well did the General regard this nephew's keenness that in 1858, when Lammot was only twenty-seven, he was dispatched to Europe to observe and report on what advances in the art of powder-making the Crimean War had brought out. But, at home, Lammot had tackled a problem that, if successfully solved, would be worth any profits of a dozen wars.

Black powder's chief component was saltpeter, which refined is almost pure potassium nitrate. This substance made up three-fourths the content of good sporting and rifle powder and about two-thirds the content of blasting powder. Since the sixteenth century, India had supplied the bulk of the world's saltpeter and its price at Calcutta, plus freight, fixed the final price of powder wherever it was made.

In 1809, vast beds of a new type of saltpeter were discovered in Peru.§ The discovery, at first, had seemed to signal the end of India's monopoly and the opening of a cheaper source of supply. But, unlike the India nitrate, which is a salt of potash, that of the Peruvian beds is a salt of soda, or sodium nitrate. Moreover, the latter contains impurities difficult to remove. Powder made with it soon became damp and refused to fire. For more than forty years this difficulty had forced the industry to con-

§Peru ceded the district to Chile in 1881, after a war.

tinue to look to India's dwindling and expensive potassium nitrate supply, despite Peru's great sodium nitrate stores.

General Henry du Pont was still immersed in the Crimean War business when one day Lammot entered his office with the air of one who had discovered gold. And gold indeed, he had found. Lammot had perfected a new blasting powder that acted favorably under the severest field tests, although it contained not a grain of India's precious potassium nitrate. Its formula was 72 parts of refined Peruvian sodium nitrate, 12 parts of sulphur, and 16 parts of charcoal.

Lammot called the new explosive "B" blasting powder or "soda powder." On May 19th, 1857, his invention was patented. In the hands of the General's expert sales agents, it swept the coal and iron fields almost overnight. Containing a higher percentage of oxygen and nitrogen, the soda powder was more powerful than most of the blasting powders on the market. It gave cleaner combustion. Glazed with graphite, another innovation in blasting powder, it poured more freely. And the cheapest and most inferior potassium nitrate powders could not compete with it in price, in a market where price ruled.

Long dominant in the manufacture of gunpowders, Du Pont now also dominated the blasting powder field. So large became the demand for soda powder that the mills on Brandywine Creek could no longer meet it. General Henry bought outright the powder mill of Parrish, Silver & Company on Big Wapwallopen Creek, Luzerne County, Pennsylvania, in the heart of the anthracite region. Rebuilt under Lammot's direction, the new mill was soon turning out 36,000 kegs of soda powder yearly, a production that was to double, and then double again.

Soda powder was the first really notable change in the composition of black powder in more than 600 years. Also it was the first distinctly industrial explosive that could not be used successfully in firearms, or practically in warfare. Before it, all blasting powders had been simply inferior grades of gunpowder. By his invention, therefore, Lammot opened a new branch of explosive manufacture devoted exclusively to highly specialized industrial needs. Thus began, indeed, a revolution that was to metamorphose the powder industry.

Soon soda powder supplanted potash powder in all kinds of blasting, except a few types of quarry work. The eyes of every nation turned speculatively toward the distant west coast of South America. There a narrow strip of land about 200 miles long, whereon not a green thing grew, suddenly became a key spot of the civilized world. That arid strip was the one plentiful source of a cheap but vital raw material upon which civilized progress during the next half century was to grow increasingly and, in time, alarmingly dependent.

Lammot's invention could not have been better timed. India's saltpeter deposits had been worked steadily for nearly three centuries. They had been heavily drained. Presently, the unprecedented needs of the American Civil War were to threaten them with exhaustion and send India saltpeter prices sharply upward.

All was not triumph for the Du Pont partners, however, in that eventful year of 1857. The country plunged again into business panic, a backwash of the Crimean War prosperity, territorial expansion and frenzied speculation loosed eastward by the yellow flood of California's gold. The General toiled over the firm's affairs as banks closed, old business houses crashed, and agents and customers joined in the common cry for credit extensions. And while his trio of candles flickered in the headquarters' office, a second light gleamed back from Lammot's chemical laboratory. The older man toiled to achieve system, sales, yet greater economy in the existing order. Lammot, with restless urge, worked to uproot the old order and establish a better one. Already he saw beyond soda powder, beyond even black powder itself.

Late August brought tragedy again. One of the older mills was being dismantled in the Hagley Yard of the Brandywine Works. Alexis du Pont was in personal charge of moving out the heavy powder-coated bins and machinery. His hands and face were sweat-streaked and black with the fine dust, as were those of his men. Past forty-one now, Alexis had been a powderman for twenty-two years. This was his life, these men were his daily companions.

A large bin was being moved. Alexis and Edward Hurst, the yard foreman, both lent a hand. They had the clumsy bin almost outside when a scuffing shoe probably emitted a spark. The thin soot-like layer of dust on the floor ignited. Instantly Alexis' clothes were in flames, but he was quick enough to leap through the door in advance of an explosion that wrecked the building, killed two workmen and fatally injured Foreman Hurst.

Dashing to the mill race, Alexis thew himself into the water. He was badly burned. The roof of a building near-by had begun to spit tiny tongues of flame, started by debris of the first mill. Shouting to the men to get back, Alexis ran to the new danger point, dripping water. He was climbing onto the roof when the building exploded under him. His men picked him up alive, but he was fatally injured.

The day was Saturday. Knowing that he was to die, the injured man spent the time left him surrounded by his family and his friends. A devout Christian, he enjoined each of them to give faithful service to God. Sunday morning he asked how long he had yet to live. He was told a very short time. He asked that the men from the mills be admitted.

A group was just outside. They pressed into his room— John Davis, Henry Danby, William Dougherty, Thomas McCallister, Hugh Stirling, John Peoples, Robert Wilson, Jonas Miller—still they came, the "Toms" and "Bobs" and "Jacks" and "Hugheys" amid whom Alexis had spent most of his adult life. They stood in the hall, and on the stairway, and in the yard.

"What a blessing," said the dying powderman, "to die with all your friends around you!"

Finally, the doctors closed the doors. He fell into a quiet sleep, from which he never awoke. Men from the mills dug his grave in the family's burial ground on the hill, and carried him there.

He left three sons and four daughters, and two enduring monuments to his memory. He founded the Cathedral Church of St. John in Wilmington, present seat of the Episcopal Bishop of Delaware, and was a leader with three of his sisters in the founding of Christ Episcopal Church, in Christiana Hundred, from which he was buried.

Lammot du Pont succeeded to Alexis' vacated place in the partnership.

Civil War

DELAWARE, by sentiment, tradition and much of the blood that flowed in the veins of its people, was bound to the South. In the tense election of 1860, Lincoln received less than one-fourth of Delaware's popular vote. Breckinridge, candidate of the Southern extremists, carried the State overwhelmingly.

The defection of the State might have counted little in the outcome of the Civil War, but in the Du Pont powder mills Delaware had an asset that weighed heavily in the scales by which were measured the forces making for victory or defeat.

Those mills constituted more than one-third of the gunpowder-producing capacity of the nation. They were the largest and finest equipped, the ablest manned. Lammot du Pont, though only thirty years old, was the nation's leading authority on explosives chemistry. As noted earlier, he was familiar with advances made in powder manufacture as an outgrowth of the Crimean War. This was at a time when radical changes were in progress in military powders and firearms, which threatened important innovations in their manufacture and use.

The revolver had been invented in 1836 by Colt. The principle of rifling the inside of a gun barrel, long known, was at last being applied to increase the range and accuracy of firearms over the old smooth-bore weapons. Metallic cartridges, breech-loading, magazine rifles and even machine-guns were under feverish experiment and were potential developments that made the trained technician a factor of stupendous importance to either side.

Against these actual or pending changes in guns stood the fact that gunpowder itself had undergone no corresponding improvement in many years, and that until new powders were compounded to fit the new firearms with satisfactory accuracy, the advances made by the arms designers would be only partially realized. In other words, the powdermaker had become the key to the situation. That fact was made strikingly clear by a condition that existed in the major classes of artillery.

In 1847, Captain Thomas J. Rodman, U.S.A., had invented the Rodman gun, a new type of cannon cast about a hollow core through which was run a cooling stream of water. The invention made possible for the first time the building of big guns with bores of 14, 15 and later 20 inches. Such guns were monsters compared to the largest in previous use, but, with existing cannon powders, were ineffective except at short range. When loaded with powder charges sufficient to propel their heavy projectiles at long range, the guns burst. Similar accidents occurred with rifled cannon because the shots, more closely fitted in the barrel, set up a greater resistance to discharge.

The difficulty was that the existing powders exerted a *blasting* instead of a *propelling* force. As a step toward overcoming this, Rodman, in 1857, invented a pressure gauge by which the force of the explosion in the gun could be accurately measured. He hoped, by changes in the powder, to control the breech pressure to precise limits. About the same time, a simple means was devised to measure the velocity of the speeding projectile.

Thus overnight, as it were, instruments became available by which guesswork in powder-making was supplanted by an exact and complicated science. The problem of the powder chemist now was to produce a progressively burning propellant that would apply its energy throughout the barrel and not suddenly in the breech. Working with Lammot du Pont, the inventive Rodman had set out to solve this problem also and, as one step forward, developed what he called Mammoth powder.

The "grains" of Mammoth powder ranged from the size of walnuts to that of modern baseballs, according to the gun caliber, and were compressed under 20,000 pounds pressure to a uniform texture and hardness. Their slower rate of combustion allowed the energy developed to take the line of least resistance, that of the open barrel. Thus the pressure within the gun's breech was reduced by one-half and the velocity of the projectile and its range increased. At last, the big gun of long range had become a formidable actuality, in its laboratory stage at least.

Also the line of division between military and industrial or blasting powders, first drawn by Lammot du Pont's soda powder, had been made even more marked. On one side of the line, hereafter were to be the *propellants,* designed only for use in firearms, and on the other side were to be the *disruptants,* suited only for blasting or shattering. All future explosives were to fall into one or the other of these two classes.

Before Lammot had time to adapt Rodman's new powder to large-scale manufacture, necessitating substitution of power presses for hand presses, the Confederates fired on Fort Sumter and Lincoln's call went out for 75,000 volunteers. War was on, with Lammot du Pont the one powdermaker intimately informed on Rodman's work and who also possessed the technical ability and plant resources necessary to put Mammoth powder, and through it longer-range big guns, to battle use.

This unique position of the Du Pont Company was emphasized by the character and location of other powder-mills in the country in April, 1861. The North had all but two of these mills. One of the two exceptions had been built in South Carolina to supply powder for blasting a tunnel, but it was too small for military use. A second mill near Nashville, although hurriedly enlarged, was capable of producing only 500 pounds of powder a day, a trivial amount for a major war.

Gunpowder and technicians who could manufacture it were among the Confederacy's most urgent needs. Shortly the South was to pay an average price of $3 per pound, more than ten times its usual cost, for English-made powder smuggled by blockade runners. By April, 1862, she was to produce an excellent grade of powder at a Government plant built at Augusta, but at no time between

1861 and 1865 was the South able to supply her infantry with more than 90 rounds of ammunition per man, whereas the Ordnance Manual called for 200.*

Above the Pennsylvania line were a dozen powder-mills on which the Federal authorities could draw, but the North's powder situation, too, was far from satisfactory, if the Du Pont mills in doubtful Delaware were omitted. Most of the establishments were small, uncertain in performance, deficient in resources. The loyalty of the one really large producer among them, the Hazard Powder Company of Connecticut, was questioned. Colonel Augustus G. Hazard, head and chief owner of the Hazard company, had lived in the South and had publicly attacked Abolitionist activities. Hazard did not want to fight the South. To complete the picture, neither did the Du Ponts.

General Henry du Pont had been, up to 1860, an ardent Whig, and held with that party's leaders, Clay and Webster, who believed that compromise should settle the difficulties between the States. He had been a delegate in 1844 to the Whig convention that had nominated Clay for the presidency. When, in 1860, the election either of Lincoln or Breckinridge meant almost certain strife, General du Pont had vigorously supported John Bell of Tennessee, the Constitutional Union candidate.

However, General du Pont dispelled all doubts as to where the Du Pont Company stood two days after the fall of Fort Sumter. He wrote the company's Richmond

*Notes of Brigadier-General Josiah Gorgas, Chief of Ordnance, C.S.A., Army Ordnance, March-April, 1936.

agent, who had just filed a heavy powder order for Virginia, that "a new state of affairs has arisen." Emphatically, he wrote:

"Presuming that Virginia will do her whole duty in this great emergency and will be loyal to the Union, we shall prepare the powder, but with the understanding that should general expectation be disappointed and Virginia, by any misfortune, assume an attitude hostile to the United States, we shall be absolved from any obligation to furnish the order."

Southern agents were recalled, Southern orders canceled, powder in Southern magazines was written off the books. Orders were dispatched to Northern and Western agents to sell not a pound of powder to buyers who might ship it surreptitiously into seceding States, which were bidding for powder at almost any price. In a conference at Washington, President Lincoln was given the company's pledge of support.

What Union loyalty meant to Du Pont, to Hazard, and every other Northern powdermaker of honest purpose was the immediate abandonment of a profitable commercial business to less patriotic competitors and to new small mills that sprang up. Also it meant expenditures, to the limit of their resources, for additional buildings, machinery and special equipment, which in a few months might be left standing idle.

The Du Pont Company was confronted by a condition that well might have awed even its hard-headed West Point-trained chief. The State boiled with secessionist sentiment. In two of its three counties, Sussex and Kent, feeling was dominant that the Southern States should be permitted to secede peaceably. The Legislature refused to vote secession, but also refused to vote loyalty to the Union. In the upper county, New Castle, Union sentiment prevailed. Wilmington bought 800 rifles and formed a regiment for defense. The city was in a vulnerable position. The Du Pont mills were only a few miles away, and through Wilmington's center ran the principal telegraph and the chief railroad line of the North, over which troops and war supplies were soon moving southward.

Into this situation General du Pont stepped officially on May 11th, when Governor William Burton commissioned him major-general in command of all Delaware troops "raised or to be raised." On that day, North Carolina was still wavering between loyalty and rebellion, and Virginia was fighting out the same issue at the polls. Five days earlier, as an incidental occurrence, the General's eldest son, Henry Algernon du Pont, had been graduated by West Point with the highest honors of his class and commissioned a First Lieutenant of United States Artillery.

General du Pont's first order was that every officer and man of the Delaware home forces should take the oath of allegiance to the United States. Simultaneously, Governor Burton directed that all arms and other military equipment be turned over to General du Pont for "inventory." Delaware was torn between approval and indignation.

Dover, the capital, was stormed with protests. Threats were hurled. Burton weakened, and, on May 14th, revoked his arms order.

Promptly, General du Pont reported to General John A. Dix, Federal commander at Baltimore, that many officers and men of Delaware's armed forces refused to swear allegiance. He asked that national troops be sent into the State. Shortly the drums and tramp of blue-clad columns sounded in Wilmington's streets and a Union detachment occupied the fort on Peapatch Island in the Delaware River. The action ended all doubt about Delaware. It was secured for the Union. Between Sumter and Appomattox, the State met every Federal tax assessment and gave 13,651 fighting men to the Union, or almost one out of every eight of her combined white and negro population.

Meanwhile, Lammot and Irénée du Pont were busy at the mills. Expansion was begun. Two infantry companies of workmen were organized and equipped. They drilled nightly. Guards were posted to keep strangers out of the Brandywine Valley. A railroad spur was built from near-by Montchanin into Wilmington to connect with the main line, to give a shorter haul to the picturesque old powder-wagons, slow and easily ambushed.

As the first year of war neared its close, a serious situation developed for the North. The fighting that had taken place, although not much more than skirmishes compared with what was ahead, had all but depleted the North's supply of India saltpeter. Between the North and India stood England, which had recognized the Confederacy as a belligerent.

At a tense meeting in Washington of Lincoln's principal advisers, Lammot du Pont was directed to sail at once for England to buy all the India saltpeter he could acquire. His mission was to be kept secret. To avoid publicity, the transaction was to be in the name of the Du Pont Company and through its regular London agents, but the United States was to be the real buyer.

On November 19th, 1861, Lammot reached London. In a little more than a week he bought about 2,000 tons of saltpeter, and had four ships waiting to load at London, Liverpool and Greenock. Part of the saltpeter was aboard them by the 28th, when a report reached England that aroused the British. The British mail ship *Trent* had been stopped out of Havana by the U.S.S. *San Jacinto,* which had fired two shots across her bow. Two of the *Trent's* passengers, John Slidell of Louisiana and James M. Mason of Virginia, had been seized and transferred to the warship as prisoners. That Slidell and Mason were admittedly Confederate commissioners en route to Europe in an attempt to enlist French and English aid against the United States did not alter the fact that their seizure was a breach of international law.

Official edict stopped the loading of Lammot's saltpeter. Armed guards were placed over his ships. English newspapers threatened war and 8,000 English soldiers set sail for Canada. Lammot took ship for America and was back in Washington the day after Christmas. Three days earlier, he learned, the British minister had served a written demand on the State Department for an immediate surrender of the Confederate commissioners and an apology for their seizure.

The demand gave the United States seven days in which to answer. However, although the ultimatum had been delivered officially on December 23rd, President Lincoln and Secretary of State Seward had been informed confidentially of it on December 18th.† In the interval, the North had become as aroused as was England herself. A banquet had been given in Boston in honor of the captain of the *San Jacinto*. Congress had voted him the country's thanks.

Although most historians ignore it, the Union was embarrassed by the war going on within its own borders. The saltpeter supply was sufficient to last only a few weeks. Without saltpeter it was impossible to make gunpowder. The South had saltpeter deposits in the limestone caverns of Tennessee, Alabama, Georgia, Arkansas and Texas,‡ but the North had none unless it resorted to compost heaps, requiring a year or two to develop. France was almost openly hostile to the North, and was doing her utmost to line up England in joint recognition of the Confederacy as an independent nation.

There is no record to indicate what happened upon Lammot du Pont's arrival in Washington on December

†Rhodes' *History of the Civil War,* Macmillan, 1917.

‡According to the Notes of Brigadier-General Josiah Gorgas, Confederate Chief of Ordnance, these caverns supplied one-half of the saltpeter consumed by the Confederacy, the other half being brought in by blockade runners from English sources.

26th. However, four days later Slidell and Mason were released. On January 1st, 1862, they took an English ship for Europe and, by coincidence or otherwise, Du Pont sailed by another ship for London on the same day armed with a letter from Secretary of State Seward to Charles Francis Adams, United States Minister to England, who was instructed to do all in his power "for the relief of E. I. du Pont de Nemours & Company."

On January 18th, the embargo on saltpeter was removed, the loading of the ships was resumed, and, the political crisis having passed, Lammot arranged to sell in England some of the saltpeter he had bought in the name of his firm. The ships sailed on February 2nd, bearing cargoes for which the United States had paid almost £80,000 and which were to make possible the continued prosecution of the war by the North, at least for another year.

The incident awoke thinking people of the nation to the urgent need of a saltpeter supply independent of any European power.

About this time another young Du Pont—Eugene, the eldest son of Alexis—joined his uncle and two cousins at the Brandywine mills. Eugene was twenty-one, a graduate chemist of the University of Pennsylvania. Lammot soon had him absorbed in laboratory work. Samples of captured Confederate powders and of new European powders obtained by Federal agents were sent to the Du Pont laboratory to be analyzed and appraised. Frequent conferences were held with military experts in a common effort to develop more effective powders and cartridges and faster ways of loading big guns.

Lammot devised iron plates with rounded hollows that looked like huge muffin pans between which the ball-size grains of Rodman's Mammoth powder could be pressed out in quantity. Federal gunners supplied with this powder were able to fire from ranges beyond those of Confederate batteries. Mammoth powder was also adapted to the large-bore Dahlgren guns of the Navy. It was used by the *Monitor* in its epochal battle with the *Merrimac* at Hampton Roads on March 9th, 1862, which marked the doom of wooden ships in war.

What effect this intensive technical work had on the war's result is problematical. Although developments were passed on to all mills supplying powder to the Union, the powder of some of these mills was notoriously inferior. The Ordnance Reports of 1864 indicate that only one plant besides Du Pont, and that comparatively small, could be depended upon to produce powders uniformly of the highest grade. However, it is coincident that as the war progressed and the Du Pont mills became more and more the chief reliance of the North, Union artillery performance improved and was probably superior to the Confederacy's during the final two years of the conflict.§

Technical skill solved the pressing saltpeter problem of the Union. By a chemical process outlined by E. I. du Pont as early as 1831, Peru's comparatively cheap and plentiful sodium nitrate was converted into potassium nitrate

§Rhodes' *History of the Civil War.*

perfectly suitable for gunpowder. India's ancient saltpeter monopoly ended, Peru's rose in the sun. Coming when refined India saltpeter cost 22½ cents a pound, to rise later to 30½ cents with the exchange rate at 13 dollars to the English pound, this development not only saved the North large sums but added to the nation's military security for more than fifty years to come.

Between 1861 and 1865, the general price index for all commodities in the North rose almost 117 per cent despite the large-scale introduction of machinery. Pig iron, quoted in 1860 in New York City at $20 to $27 per ton, sold at $43 to $80 per ton in 1864. Other badly needed basic materials soared correspondingly. In May, 1864, General du Pont wrote that saltpeter had advanced 135 per cent in three years, sulphur 80 per cent, charcoal 50 per cent, cooper work 90 per cent and Du Pont wages 75 per cent. The wage rise compared with one of 43.1 per cent for labor generally.‖

Yet only during the closing months of the war did the price of Du Pont powder sold the Government reach 33½ cents a pound, a figure that was but 8½ cents higher than the price of first-grade powder in 1800, more than sixty years before. In December, 1861, with powder quoted at 20 cents in the market, Du Pont's price to the Union was 18 cents, or less than the peacetime average.

Mounting costs and heavy taxation forced the price up to 26 cents in 1863, which again was less than the market price. At that time, a War Department inquiry revealed,

‖*Statistical Abstracts,* 1899.

England was paying 34 cents a pound for cannon powder and 40 cents a pound for musket powder, although wages were much lower in England and both saltpeter and sulphur were brought in duty free and at lower first costs. Powder made in the non-profit Confederate mills at Augusta cost $1.08 a pound during the first year of operation, when material costs were less than at any subsequent time.¶

Roughly estimated, Du Pont furnished between 3,500,-000 and 4,000,000 pounds of military powders of all kinds to the Union Army and Navy during the war. This total exceeded probably by a million pounds that of the Augusta plant of the Confederacy. This was a record in which General Henry took a personal pride.

"There never has been a case in any country in the world," he wrote in 1864, "where a nation at war has had its powder so cheaply as the United States have had it since the breaking out of the Rebellion." And on the profit side of the Du Pont ledger he noted that the hectic wartime business had not equaled "the regular demand which would have existed had peace continued."

Southern sales had been lost. New competitors dotted the coal and iron fields. In California, the gold miners, unable to get blasting powder from the East because of the menace of Confederate privateers on the sea, had raised $100,000 and built the California Powder Works,

¶*History of the Explosives Industry in America,* by Arthur P. Van Gelder and Hugo Schlatter (Columbia University Press, 1927).

which in 1865 were manufacturing more than 500,000 pounds of blasting powder a month. This lost bonanza, which the General himself had fostered, made him boil in anger whenever anybody spoke of the "fat profits of war."

At the front, the General's son, Henry A., rendered a service no less distinguished than his record at West Point. Promoted to a captaincy early in 1864, he commanded Light Battery B of the Fifth United States Artillery at the battle of New Market so well that he was made Chief of Artillery of the Department of West Virginia. In September, he was brevetted a major for "gallant and meritorious conduct at the battles of Winchester and Fisher's Hill." One month later, he was raised to lieutenant-colonel for "distinguished services at the battle of Cedar Creek," objective of Sheridan's famous ride.

The highest honor the United States can confer on a soldier for heroism is the Congressional Medal of Honor. That medal was awarded Henry A.

Another Du Pont won recognition. Samuel Francis, a son of Victor and graduate of Annapolis, was a captain in the Navy when the war began. Appointed a flag-officer and president of the Naval Strategy Board, in November, 1861, he led out of Norfolk a Union fleet of seventy warships and transports loaded with soldiers, the greatest ever assembled under the American flag. Attacking Port Royal, a fortified island off the South Carolina coast, Du Pont captured it and established a Union base, from which he was able to clear the coast of Confederate ships south to Florida. The action established the blockade of the Confederacy.

Du Pont was thanked by Congress and made a rear-admiral. Du Pont Circle in the nation's capital was named in his honor, as was also Fort du Pont in Delaware. He died before the war ended, broken-hearted over the failure of his last great naval fight in Charleston Harbor, which, he felt, was lost because the Army had not heeded his request to cooperate in the action.

With peace, the Government asked to be released from all powder contracts yet unfilled. General Henry complied unconditionally. Few of the other powder companies were so disposed. They insisted the Government either buy their unfilled contracts outright or accept the powder it had ordered. Washington decided to accept the powder.

Tons of unwanted powder were dumped into Federal arsenals, already bulging with war surpluses. Confederate surpluses added to the war's leftovers. It seemed a smart move by the powdermakers. Their contracts called for powder at 30 cents and 33½ cents a pound, the war's highest prices. Raw material costs had dropped and wages with them. Crude saltpeter prices, for instance, had declined in a few weeks from 25½ cents a pound to 12½ cents, which allowed fat additions to profits that Du Pont influence during the war had kept lean.

Powder-swamped and debt-troubled, the Government struck back early in 1866. The powder that was being dumped into the arsenals was dumped onto the market.

Public auctions were held. The powder was sold for cash to the highest bidder, bringing as little at 5 cents a pound. Individuals went into the powder business overnight with only a desk and chair for equipment. Shortly the whole trade was disorganized and fighting for its life.

In exasperation, General du Pont offered to contract for all the Army powder at a fair price and release it gradually into the market, as had been done after other wars. No official dared take up the offer. Discharged soldiers had come home with accounts of wormy beef, faulty rifles, shoddy blankets and paper shoes. The nation was rumbling angrily over fraud and graft in war contracts. Any new contract with a powder company would raise suspicions, regardless of how much money it might save the public purse, officials explained.

Customers demanded the cheap Government powder. Against his will the General was forced to buy at the auctions. Here the powder was sold in barrels without opportunity of inspection. His temper was not improved, when, in a purchase at Charlotte, his agent paid for first-grade rifle powder and got "a general assortment of percussion caps, fuses, ends of rope, old nails, spikes, paper and brass balls." But much of the powder so sold was good.

The public powder auctions continued intermittently for six years, then were followed by private sales. Until 1890, re-made Civil War powder was still on the market.

*

BOOK TWO

*

High Explosives

A King Totters

BLACK powder was revealed to be woefully lacking as an all-round explosive during the rush and development in the mid-century. It had shortcomings in mining. It pushed rock or ore from its bed in huge chunks, which then had to be broken into pieces. This multiplied costs and retarded work when the need was for greater speed. What industry required was a blasting agent that not only would push but also shatter, whole mountains if need be.

Two noteworthy discoveries had been made in 1845. At the University of Basle, Switzerland, Professor Christian Frederick Schoenbein produced a nitrated cotton which he called guncotton, and at the University of Turin, Italy, Professor Ascanio Sobrero invented nitroglycerin.

No other one business eventually was to undergo greater changes than was the Du Pont Company as a consequence of these two discoveries. In modified forms, the two literally were to alter also, in time, the face of the earth.

Both explosives were smokeless, exploded cleanly and had tremendous power. At once newspapers hailed guncotton as the successor to black powder, but the careful Alfred du Pont, then head of the business, was not so sure.

"The discovery is brilliant and such as to create astonishment," was his verdict, "but the introduction of guncotton in common use must be the work of time."

The Du Pont Company was to rest on that verdict for more than forty years.

More daring and impatient, a company of Englishmen built a guncotton plant at Faversham, England. In 1847, it was destroyed with so heavy a loss of life and property that no English manufacturer touched guncotton again for almost two decades. Russia, France, Germany and Austria experienced similar tragedies.

Meanwhile, Sobrero, appalled by the might and deadliness of his own discovery, warned the world against attempting to convert nitroglycerin to industrial uses. It was more treacherous than guncotton, he reported. Sometimes the slightest jolt would detonate it, again it would withstand a hard blow. Slow heating caused it, seemingly, only to burn; rapid heating, to explode. Sobrero said he could establish no rules to govern it.

But, in 1863, a startling report came out of Sweden. There the inventor, Emmanuel Nobel, and his son, Alfred, dared to put the might of nitroglycerin to test in blasting granite. Results were described as astounding. Next came word that Alfred Nobel had invented a device by which nitroglycerin could be detonated surely and in comparative safety, if it was confined.

Nobel had contrived a small copper cylinder loaded with black powder—later he used fulminate of mercury— to serve as a detonator. This he placed in the bore hole next to the explosive charge and fired it with a fuse.

Known today, vastly improved, as the blasting cap, Nobel's device was revolutionary. It enabled man for the first time to release at will the terrific energy imprisoned in high explosives. It made their daily use in industry a practical possibility. Such was the situation as the Civil War ended and America turned again to the growing industrial problems brought by peace.

Shortly, nitroglycerin was employed in its first large-scale industrial operation, the blasting of a railway tunnel through solid rock near Stockholm. Plants for its manufacture were built in Sweden and Germany. December, 1865, found Nobel's "blasting oil," as it was called, thundering its success in a dozen countries, including the United States of America.

Sobrero's grim forebodings were forgotten. Nobel's factories hummed. They sealed the oil in zinc cans, packed the cans into wooden shipping crates, stuffed the crates with wood shavings—later with kieselguhr, a porous clay found in abundance in northern Germany—and nitroglycerin traveled the highways of the world. Peasants in Europe's mountains jolted the liquid death over rough roads in two-wheel ox carts.* When the cans leaked, which they often did, the peasants rubbed the thick oozing oil, in lieu of axle grease, on their boots. Railroads and ships carried the oil as casually as if it were lard. Salesmen called on prospective customers with bottles of the stuff in their carpetbags.

"All hell will be to pay for this! Wait!," General Henry

*Nobel, Dynamite and Peace, Ragnar Sohlmann and Henrik Shück, Cosmopolitan Book Corporation, 1929.

du Pont warned American quarrymen and miners who brought glowing accounts of the mighty blaster that made the Brandywine's best black powder look puny.

The wait was not long. A negro porter in the Wyoming Hotel of New York City's Greenwich Village noticed that a reddish smoke was coming from the box he had been using as a seat for his boot-blacking stand. The box had been checked at the hotel by a German traveler, who had neglected to mention that it contained nitroglycerin. The porter put the smoking box in the street and ran. A moment later it exploded, ripping a hole in the stone pavement and seriously damaging the hotel and surrounding houses.

That was the prelude. On March 4th, 1866, two cases of nitroglycerin exploded in Sydney, Australia, demolishing a warehouse and other buildings with a heavy loss of life. A month later the steamship *European,* with seventy of Nobel's wooden-crated cans in her hold, together with ammunition for South American revolutionists, blew up while unloading at Aspinwall, now Colón, Panama. When the debris was cleared away, sixty dead were counted. Property damage exceeded $1,000,000.

Only a few days passed when the canned death rocked the world again. A shipment of it destined for the California gold fields rent the Wells Fargo Express offices in San Francisco and killed fourteen, injured more. In May, Nobel's factory in Germany was wrecked by a blast. Still later, ten persons were killed near Brussels, Belgium.

A horrified world cried halt. Laws were enacted restricting or prohibiting the manufacture and shipment of nitroglycerin. Transportation companies refused to handle it. Nobel's agents dumped their stocks into rivers or the sea. Before his chemistry classes at Turin, Ascanio Sobrero bowed his handsome head in shame that he had discovered this destroyer.

Yet the fact stood that nitroglycerin, with all of its faults, would do almost five times the work of black blasting powder and do it better. Properly controlled, it was a builder mightier than any the world had known. Tame it, harness its energy, and it might emancipate mankind from its most back-breaking tasks. Nobel, the semi-invalid, outcast even among his friends, faced the storm with this fact blotting out all others. What of black powder's record, he asked; it, too, was bloodstained. Back he went to his laboratory.

As early as 1863, he had conceived the idea of reducing the sensitiveness of nitroglycerin by having it absorbed into a porous insensitive substance as water is absorbed by a sponge. He revived this idea. One after another he experimented with powdered charcoal, sawdust, cement, brick-dust, and finally powdered earth—kieselguhr—already employed to cushion his zinc cans in their wooden crates. He found kieselguhr would absorb three times its weight of nitroglycerin. Moreover, when so saturated, it became a putty-like substance that could be kneaded and packed in cartridges, which were surprisingly less sensitive to shocks or blows but could be exploded with a blasting cap.

With eager fingers, Nobel kneaded up a batch of the putty into short round sticks that would readily fit into bore holes for blasting. He wrapped the sticks in tough waxed cartridge paper, packed a box with his new blasting caps and set out for the German mines. His older brother, Robert, took a similar kit into the stone quarries. All summer they blasted. Results in shattering hard rock showed that although the power of the new explosive was less than that of straight nitroglycerin, its violence, or "brisance," was more than twenty times that of black powder.

They set fire to the oil-soaked sticks and watched them burn harmlessly, whereas even a tiny spark would explode black powder! They hurled boxfuls of the sticks from cliffs onto jagged rocks, but nothing disastrous happened. By October, 1866, so confident were they of the comparative safety of the kieselguhr-cushioned nitroglycerin under all reasonable conditions of use that they invited the leading European experts to Germany to make any tests they wished. When the experts had finished, that year of disaster became also the year of Alfred Nobel's greatest triumph. His invention—dynamite—was the nearest man had come to creating a manageable mountain-wrecker!

Soon builders were calling dynamite one of the greatest boons to progress since the invention of the printing-press. It was to do for mining and construction what steam-power was already doing for transportation and manufacturing. Before long, too, it was to turn inside out the Du Pont plant on the Brandywine, which in 1866 was the largest producer of black powder on five continents.

Powder Trust

LAMMOT DU PONT had studied Sobrero's work. He had followed Nobel's difficulties with nitroglycerin, secretly had conducted experiments of his own that had convinced him not only of dynamite's practicability but also of its enormous field in America. Developments now confirmed his conclusions.

The Giant Powder Company, the first dynamite plant in America, was organized in California in 1868. The plant's crew was only a nitroglycerin maker, an assistant, and a few Chinese laborers, but soon gold miners were using dynamite at $1.75 a pound; it was cheaper than blasting powder at 20 cents a pound. Gold was being washed out of mountain sides by the rush of water piped from high levels. When the water struck hard rock it was necessary to blast. A tunnel would be cut, loaded with 300 to 700 kegs of black powder, which would heave the rock out in chunks the size of a miner's cabin. A fourth as much dynamite got rid of the rock in a fraction of the time.

Over the line in Nevada lay the great Comstock lode, with its mine shafts striking deep into the scarred face of Mount Davidson. By the end of the Civil War, the Comstock mines were producing silver bullion with a coin value exceeding $15,000,000 per annum. Dynamite, instead of black powder, went down the Comstock shafts and the quantity of silver coming up more than doubled. The output in 1877 reached $38,000,000, despite difficulties in mining previously regarded as insurmountable.

Black powder had been hacking away at the rocky base of the Hoosac Mountains near North Adams, Massachusetts, in an effort to cut a six-mile railway tunnel. At the end of a mile, the work had been halted by solid rock. With nitroglycerin the tunnel was completed without a serious accident. It set a precedent for future tunnel-building.

Soon dynamite plants dotted the country. Humble affairs most of them, often housed only in farm buildings, they were the pioneers of a new industry that grew in spite of all obstacles. Explosions, deaths, failures, frauds, lawsuits, financial panic, prejudice, the full weight of the black powder industry's opposition—all could not check it. New and better plants rose from the debris of old ones, new and more daring feet stepped into dead men's boots. When railroads refused to carry dynamite, iron-nerved salesmen recruited from mines and quarries packed it in trunks as "personal baggage" or in boxes labeled *"Glass, Handle with Care."* It was stored in hotel sample rooms as *"Chinaware, Fragile,"* hidden in cellars, under beds. Demonstrators, with Colt "six-guns" on their hips, lugged it in suitcases into hostile mining towns, held it between their fingers as it burned, pounded it on planks with sledge-hammers, set off blasts of it with the lighted tip of their cigar touched to the fuse of a detonating cap.

When they could not sell the product they gave it away for trial to sell itself.

And dynamite at $1, then 75 cents and then at 50 cents a pound, as the facilities for producing it grew, swept *via* Great Lakes freighters into the copper and iron districts of Lake Superior. River boats carried it to St. Louis to be hauled overland to the lead and zinc mines of Joplin. It dug sewers in New York, uprooted stumps in Ohio, drained swamps in Louisiana, began furnishing millions of tons of limestone for flux essential in smelting iron ore and in making steel.

History records an extraordinary American industrial advance following the Civil War. Between 1870 and 1880, we built 40,545 miles of railroads, or three-fourths of the total built in the previous fifty years. During the next decade, 73,410 miles of railways were laid. Coal tonnage rose from 33,000,000 tons in 1870 to over 157,000,000 tons in 1890. Production of iron quadrupled, copper output multiplied nine times, silver output more than five times. In every sector, industry seemed suddenly to have donned seven-league boots. And the new boots were dynamite!

Not a barrel of cement was produced in America until dynamite made big-scale rock blasting practicable. Most of our highways were wallows of mud or dust, according to the season. Nickel was a precious metal before dynamite; only the rich could afford silver knives, forks and spoons. Seemingly impenetrable layers of rock imprisoned the bulk of our common minerals.

As Lammot du Pont saw this development approaching, he was immediately aware of the opportunities it presented. He visualized a new era in the making. But when he went to General Henry du Pont with the proposal that the Du Ponts manufacture dynamite, he was answered by a blunt and unqualified "No!"

A variety of dynamites had appeared on the market. Some were improvements on Nobel's idea in that they substituted for the non-burning kieselguhr combinations of such "active" materials as mealed gunpowder, saltpeter, sawdust, cotton, sugar and the like, which aided in the explosion and added to its force. Others were mere evasions of the Nobel patent. At mention of any of them General Henry erupted.

"It is only a matter of time *how soon*," he wrote in 1871, "a man will lose his life who uses Hercules, Giant, Dynamite, Nitroglycerin, Averhard's Patent* or any explosive of that nature. They are all vastly more dangerous than gunpowder, and no man's life is safe who uses them."

When the Pennsylvania Railroad, in 1873, considered removing its ban on dynamite shipments, Du Pont warned it of the calamities it would invite by carrying on trains "any compounds of nitroglycerin," and added: "We have sent circulars to all of our agents cautioning them against allowing any such to be stored in our magazines."

Throughout the hectic Seventies, that decade of business shaken to its base by the crash of the great banking house of Jay Cooke and the failure of three-quarter-billion

*Ehrhardt's Powder probably was meant.

dollars worth of American mercantile establishments, black powder alone rolled from the Du Pont mills—35,000 pounds of it daily. Five-sevenths of this output was blasting powder, no less than seventeen separate grades precisely compounded to meet, in General du Pont's opinion, every conceivable need of industry. In addition, the mills turned out twenty-four varieties of gunpowders.

Watching the thousands of metal powder kegs rolling ceaselessly from the mills—the mills of his father—General Henry saw them as symbols of substantial and enduring accomplishment. They were the products of an investment of brains, blood, dollars and sweat that only now was coming into fruition. They were tangible messengers of progress, and compared with them the promises of dynamite and all its kin were merely will-o'-the-wisps to which he would sacrifice no man's life, nor let Lammot sacrifice his.

As the General's shrewd eyes followed the long rows of kegs, he also dreamed of new empire and better days undisturbed by Government powder auctions and the price-cutting of competitors. He dreamed practically, however, and out of his dreams grew a plan.

Early in 1870, through a merging of interests that operated eight black-powder plants, a powerful new factor had entered the field—the Laflin & Rand Powder Company of Esopus, New York. This, with the Hazard Powder Company of Connecticut, and Du Pont of Delaware, made three big-scale producers of black powder in the United States. Four smaller but substantial companies, plus a fringe of minor establishments, numerous since the

Civil War but chiefly local in character, constituted the remainder of the industry east of the Rocky Mountains.

The Laflin & Rand company was amply financed and ably managed. Its plants had established reputation and were successful. The Laflin family's experience with black powder dated from the Revolution. Paradoxically, however, instead of worrying General Henry, the appearance of this formidable combine was a source of satisfaction to him. Ever since 1859, when he had found that economic arrangements with other big producers were desirable, he had kept on friendly terms with important competitors. He knew the Laflin & Rand principals and was confident that the new giant of Esopus would cooperate with the Du Ponts and the Hazard company on all industry matters of mutual concern.

At the moment, one matter of concern was the competition, utterly unscrupulous and unreasonable in his opinion, of the small powder plants that had sprung up to grab the business of their localities while the regular producers were occupied with the Civil War. True, these war-born ventures had supplied blasting powder in emergency to mining areas that, otherwise, would have been neglected and so had performed a public service. But in return for this service, and poor powder in the main, they had exacted high prices as long as the war-emergency existed, only to undercut the lowest prices listed the instant the regular producers reappeared in their districts. Most of these small plants made no allowances for depreciation or replacements in the event of explosions. Situated in the districts they supplied, they had no freight bills to pay and used only the cheapest kegs for containers. Individually their competition was negligible, but collectively they harassed the industry.

Surplus war powder aggravated the situation. Moreover, practices had arisen in the industry proper that verged on the scandalous. Railroads gave secret freight rebates to powder companies from whom they bought. In turn, the powder companies granted secret price concessions to railroads. Sales agents cut prices in order to steal the customers of other agents employed by the same company. Bribery and fraud were the open sesame to countless public contracts.

No agency existed that could clean up these abuses. However, in the advent of Laflin & Rand, a third big producer upon whose aid he could rely, General Henry saw an opportunity to create such an agency and make it function by right of might. A series of negotiations followed. In April, 1872, they developed into action. The chiefs of all the important black powder companies east of the Rockies met in New York. By unanimous agreement, they formed the Gunpowder Trade Association of the United States, later to be known popularly as the "Powder Trust."

Du Pont, the Hazard company, and Laflin & Rand were allotted ten votes each in all matters for decision. The Oriental Powder Company, headquartered in Maine, was granted six votes. The American Powder Company of Massachusetts and the Austin and the Miami powder companies of Ohio were given four votes each. Lammot du Pont was elected president; A. E. Douglass of Hazard, vice-president; and Edward Greene of Laflin & Rand, secretary and treasurer.

The country was divided into districts, and scales of minimum powder prices were set for each district. Sales quotas were allocated to the seven associates, an agreement was drawn and signed that if any one of the four smaller companies exceeded its quota it would buy the powder in excess from the "Big Three." In turn, if the "Big Three" sold more than their quotas in any year, the quotas of the smaller companies were to be increased proportionately. Penalties of $1 per keg were to be assessed for every instance of price-cutting or other irregularities by the associates in gaining sales.

The plan of action was simple and direct. If non-members sold their powder at not less than the minima set for members, they would not be molested. But if they should invite business war by undercutting the minima, war they would get, with the pooled resources of the associates pitted against them. Rebating, secret price lists, "cut-throat" tactics by agents, bribery and other unsavory practices were to end. And to show that it meant business, the association bought several small plants regarded as particularly obnoxious in their business methods and dismantled them.

Peace settled over the industry. The independents did not dare challenge the "Trust," and, besides, they soon discovered that they could still get business, and, what was more, earn profits at the minimum prices established. The listed prices were fair—13 cents and 15 cents per pound for blasting powder and 25 cents for sporting or rifle powder could not be called excessive by the most penurious customer. In one respect, the advantage was with the big companies in that they could produce better powder, but, on the other hand, the small producers had lower overhead costs and they still enjoyed the advantage of no freight to pay. Everybody seemed happy.

Then, almost without warning, came the panic of 1873. The black-powder industry was hit hard. At one time, half the powder-producing machinery of the country stood idle. The industry had not only depression but dynamite to face, and dynamite kept blasting gaps in black powder sales. Call money on Wall Street reached the fantastic rate of 1½ per cent a day, so there was small help to be had from the banks. Almost a third of the country's railway mileage went into receivership, with powder companies listed heavily among the worried creditors.

The panic lasted six years. During those years General Henry du Pont, who still signed his checks with a goose quill but who had good checks to sign, more than doubled the Du Pont holdings in the powder business.

Colonel Hazard had died in 1868. He had left his fortune tied up in railroad stocks and his big powder plant at Hazardville in the hands of three men, two of them old and one ill. An explosion in 1871 had seriously shaken the company, the panic jolted it again, and by 1876 the management was in distress. General Henry bought control of

Hazard in the largest powder deal negotiated in America up to that time.

That same year the California Powder Company, war-born progeny of powder-needy gold miners, became another problem. Its black powder began spilling over the Rockies and its management refused to recognize prices dictated in the East. However, some of its chief stock-holders were growing weary of the powder business and its endless risks. General Henry's long arm, which had reached to California to supply powder in the heyday of her gold rush, now reached again. An important interest in the California company passed to Du Pont.

Southward roamed the General's eye. Machinery of the confiscated Confederate mills at Augusta had been sold at auction in 1873. Present at the sale as a buyer, but short of the required cash, had been Samuel Watson, owner of the Sycamore Mills of Tennessee, which had supplied the South's first battle-powder. A Du Pont agent at the sale had advanced money to Watson, in exchange for 500 shares of Sycamore stock. Then Watson died, and for the third time in 1876 the General signed a check that bought a powder-mill. Sycamore had a sales office in New Orleans, an output of 7,500 pounds of powder daily, and in five states customers who had not forgotten Appomattox.

One-third of the stock of the Austin Powder Company of Ohio, association member with four votes, was added to the mounting pile of certificates in the General's safe. In a series of deals, he bought eight independent mills in the Pennsylvania coal fields, built an additional plant there and extensively improved his acquisitions. Jointly with Laflin & Rand and the Du Pont-controlled Hazard company, he took over the majority stock of the Lake Superior Powder Company, bothersome independent in the copper and iron country of upper Michigan.

Finally, the Oriental Powder Company of Maine, with six votes in the association's roll-call, dropped into the General's lap. Oriental had essayed to make dynamite to its sorrow. An explosion in 1870 wiped out its plant. Sold at auction, the company had been reorganized strictly as a producer of black powder, only to be struck head-on by the panic. Its debts in 1879 were $648,000, when two-thirds of its stock passed to Du Pont and the rest to Laflin & Rand.

As the decade ended, the only serious rival left in the industry was Laflin & Rand, a company that was in fact a comrade-in-arms. Du Pont votes dominated the Gunpowder Trade Association. It was with truth that the General wrote a Texas sales agent a few years later:

"We manage our own business in every particular, and allow no trusts or combinations to rule or dictate what we shall do or what we shall not do. We make our own powder, and we make our own prices at which it shall be sold, here, there, and everywhere in the world where it is for sale. . . . We do our own dictating. . . . If we choose we can as quickly as wires can carry the orders change the price at each and every point in the world where Du Pont powder is for sale. And no trust, no combination, no set of people nor persons can interfere. We have not changed our mode of selling. Our mode today is the same as it has been since our firm was established . . . and we expect to continue a hundred years more in the same way."

Price of this dominance, additional to money paid out, was ten years of grueling labor.

The companies taken over were sick; health had to be pumped into them. Despite depression, earnings of the Brandywine mills had to be maintained to justify and help finance the outside purchases. To accomplish this dual purpose, every Du Pont was mustered into a service that knew no hours, no excuses for failures, no choice of jobs.

The late Alexis' third son, Francis G., a chemist and student of astronomy, put on a powderman's overalls in 1871. He was a member of the firm and managing a mill at twenty-four. Colonel Henry A. du Pont resigned his army commission to become sales manager and practically live "on the road." William du Pont, ninth and last of the General's family—seven of whom were daughters—returned home from the Massachusetts Institute of Technology and joined the partnership.

A new grave appeared in the cemetery on the hill. Eleuthère Irénée II, production chief for twenty-seven years, checked his last powder run in 1877, a spent man at the age of forty-eight. Except for brief and rare business trips, he had never been away from the mills that had claimed him since youth.

One day, so the story goes, a grimy-faced man in powder-smeared work clothes appeared in a Wilmington store and selected a cheap rubber coat. When he attempted to walk out with it, saying he would send the purchase price over from his bank as soon as he could cash a check, the merchant blocked the doorway.

"But you should know me. I'm Irénée du Pont," the customer explained.

"If you're a Du Pont, I'm General Grant," said the merchant, and added pointedly that the coat would have to remain in the store until it was paid for; otherwise, he would send for the police.

Sighting himself in a full-length mirror, Irénée du Pont laughed, apologized to the merchant and agreed to fetch the cash.

Such was the Du Pont road during the Seventies—a road of work, sweat, and grimy hands.

Repauno

THE black powder interests of Du Pont sprawled across the country in 1880, assembled under the genius of the General, now almost seventy years old. Earnings religiously reinvested in the business had paid for them. Work that spared no male of the line had fitted the parts into a cohesive whole.

About this time an engineer of prominence applied for a job. In reply, the General wrote: "We build our own machinery, draw our own plans, make our own patterns, and have never employed anyone to design or construct our mills or machinery, dams or races, roads or anything else; being our own engineers and superintendents of all work done at the mills."

"We" were the Du Ponts in person. The wheel-horse was Lammot—chemist, inventor, engineer, builder, gang boss, dreamer.

Lammot was now forty-nine. He wore a mustache and a thin, full, fairly short-cropped beard that was slightly streaked with gray. Years of close work in the laboratory, much of it by lamplight, had put him into steel-rimmed spectacles. He was gaunt and spare. He could nurse an experiment with the touch of a woman, or outwork and outswear the toughest powderman in the business.

At one time Lammot served as paymaster for the Wapwallopen Mill in Pennsylvania. Asleep one night in the dingy coal-town hotel where he usually stopped on these monthly visits, a sound at his door awakened him. In a traveling bag underneath the bed was the month's payroll. Lammot feigned sleep. His door opened. A dark-lantern's ray flashed in his face, around the room, finally under the bed. A man entered the room and getting on his knees beside the bed began groping for the traveling bag. In one quick swoop Lammot enveloped the fellow in the heavy bedclothes. Then he picked him up bodily, carried him into the hall and pitched him down-stairs into the lobby below. Borrowing bedclothes from an unoccupied room, he returned to bed and to sleep.

Next morning, the hotel was in confusion. The landlord had vanished during the night, so hurriedly that he had not taken even a hat. The heap of bedclothes, found under the stairs in the lobby, was recognized as coming from Lammot's room, but he merely chuckled and insisted he had slept soundly. On subsequent visits to the hotel, which Lammot continued to patronize, the landlord unfailingly vanished when he hove in sight.

It was one of Lammot's best talents, this ability to turn a fight into something at which he could laugh. He was to do it in his fight to put the Du Pont firm in the dynamite business, despite his Uncle Henry's "No!"

One morning, in 1876, when near control of the California Powder Company passed to the four-room plank-floored offices on the Brandywine, General Henry awoke to find himself in an embarrassing situation. By some un-explained hocus-pocus, which suggested the fine hand of Lammot, he had become an important factor in one of the largest dynamite-producing companies harassing the black powder industry.

The California company had been manufacturing a brand of dynamite known as "Black Hercules" since 1869. This was a simple mixture of blasting powder and nitro-glycerin that had been compounded in an effort to check the inroads into black powder sales being made by "Giant" dynamite manufactured in California under Nobel patents. Nobel, in 1863, had made a similar compound, but had not troubled himself to patent it. Miners did not think much of "Black Hercules." Neither did the General. In fact, he had denounced it.

However, James Howden, able San Francisco chemist who had been retained to meet the threat of "Giant" dynamite, had not stopped with "Black Hercules." The chief fault of Nobel's kieselguhr dynamite was that 25 per cent of its weight was an inert ingredient, a porous earth which took no part in the explosion. Howden pro-posed a dynamite that would be 100 per cent explosive. Out of his experiments came "White Hercules." This new dynamite contained 75 per cent nitroglycerin absorbed in a mixture of sugar, magnesium carbonate and potassium nitrate, or India saltpeter. It was a stronger and better explosive than kieselguhr dynamite. Patented in 1874,*

---

*The patent was applied for in the name of Joseph W. Willard (No. 157,054 of July 9, 1874).

after Howden's death, the courts held it was no infringement of the Nobel patent, inasmuch as it embodied the new principle of an active instead of an inert base.

Howden's "White Hercules" opened the way for a new category of graded active-base dynamites, ranging from relatively slow-acting explosives that contained as little as 5 per cent nitroglycerin to those of lightning-like quickness containing as much as 75 per cent nitroglycerin. Consequently, when the dynamite-hating General acquired his dominant interest in the California company, which he saw primarily as a maker of black powder, he also had responsibility for one of the most potentially valuable patents in the high explosives field. Moreover, it quickly developed that a major portion of the California company's million-dollar capital was invested, not in black powder, but in dynamite.

As head of the Du Pont firm, the General was now confronted with the necessity of a no-choice decision. The California company was too deeply involved in dynamite, and too strategically situated, to be turned back into black powder alone, except at heavy loss. The General swore feelingly, but the firm was committed, its funds were invested, facts that Lammot drove home. Grudgingly conceding that "Hercules" was probably "the best of all patent powders," the General approved its manufacture by the California company.

The action led the General straight into a second situation. The Union Pacific Railroad, the one transcontinental line that linked California with the world east of the

Rockies, suffered a series of accidents with dynamite in transit. The road refused to accept another pound of the explosive as freight, an ultimatum in which the General under other circumstances might have rejoiced. But "Hercules" dynamite was now shut out of its rapidly expanding sales territory eastward, a ruinous predicament. The only solution was erection of another dynamite plant outside of California, preferably in the Midwest. A site near Cleveland was chosen. Estimates were that "Hercules" could be produced there at a saving of 10 cents a pound under previous costs. "Giant" dynamite was already entrenched at Kenvil, near Dover, New Jersey, so again the General had no choice. He sanctioned building of the Cleveland plant.

The added producing capacity of the Cleveland plant made the California company's sales organization inadequate in the East. Du Pont black powder agents, on the other hand, were clamoring for the right to sell dynamite for purposes that black powder could not serve nearly as well. In view of their demand, it would have been folly to establish a separate "Hercules" sales force in the East and Midwest to cut further into Du Pont agents' commissions. So a third major reversal of policy was forced, without choice, upon the embattled General. He authorized the firm's agents to add "Hercules" dynamite to the line, "provided they do not store it in our magazines." At last, the greatest of black powder producers and dynamite's once most powerful foe was, ironically, in the business of manufacturing and selling dynamite in all but name!

Contemplating that fact, Lammot du Pont's eyes twinkled behind his spectacles. Handy on his desk he always kept scissors and a mirror for use in trimming his beard. It was a diversion that helped him to think. Now, long legs propped against the desk, mirror in left hand, scissors in the right, he snipped whiskers with a satisfied precision. Direct assaults, he knew, would not budge the General, but the old warrior at last had left a flank open to attack.

As president of the Gunpowder Trade Association, Lammot had established a friendship with Solomon Turck, chieftain of Laflin & Rand. In early days, Turck had driven mules to a powder-wagon. Later he had served two terms as mayor of Dubuque, Iowa, when that city was still homespun and Far West. It was Solomon Turck's boast that he had learned his political and business strategy in a stern school.

No evidence exists of what transpired when these two, Turck and Lammot, put their heads together. The record does show, however, that shortly General Henry du Pont received a remarkable letter. Signed by Turck, the letter stated that he had been instructed by his Board of Directors "politely but positively" to assert the Laflin & Rand company's claim to a common interest with "your firm and the Hazard Powder Company in the high explosives business."

The General read, sputtered, and capitulated. "We are going into the high explosives business," he announced on January 28th, 1880. "That is, we are forming a company in which we are heavily interested to manufacture the same, and have not as yet fully determined on the name."

The announcement was far from the trumpet-call of an enthusiast launching upon great adventure, but three days later Lammot's workmen, assembled with suspicious dispatch, were turning up the frozen dirt in the flat open fields near Gibbstown, New Jersey, where the Repaupo Creek empties into the Delaware River opposite Chester, Pennsylvania. The sound of picks and spades in the midwinter air signified for Lammot the *Ultima Thule* of more than a decade of undiscourageable planning.

He had won an ally in the General's youngest son, William, now twenty-five. William's chief work had been management of the firm's extensive farms. He wanted a more direct connection with the business world. Fortunately, for this stage of the venture, he was an excellent oarsman. A hired rowboat provided the only means of crossing the Delaware from Chester to the Jersey side. William rowed Lammot to and from the rising dynamite plant. He also served as construction boss.

Four months after the first pick had broken the Repaupo meadows, the first charge of nitroglycerin was run through the nitrating tanks. By way of ceremony, Lammot himself carried the first bucket of honey-like liquid to the mixing house. That week of June the first ten cases of the new dynamite, christened "Atlas," were en route to the stone quarries of Crum Creek in near-by Pennsylvania. In July, production had reached 2,000 pounds daily and was rising steadily.

"We have begun here," predicted Lammot, "what will some day be the biggest dynamite plant in America." Truthfully, he could have embraced the world in that prophecy!

Meanwhile, the Repauno Chemical Company was incorporated in Delaware, the "n" of its name being substituted for the "p" of Repaupo, because Lammot found it more euphonious. He was made president of the new corporation, and young William du Pont, secretary and treasurer. An initial capital of $300,000 was subscribed equally by the Du Pont firm, Hazard, and Laflin & Rand. This gave the Du Ponts a two-thirds control.

The time was auspicious. The country had shaken itself free from the post-war mire. Gold payments had been resumed. Business indices rose in a prosperity that was to continue for nearly fifteen years.

Repauno was hardly well started before it had to expand. New land was acquired and new buildings added, but demand still exceeded capacity. On "the Brandywine," General Henry worked by candlelight and laboriously wrote his letters by hand, but at Repauno the newest was none too modern for the aggressive Lammot. He installed the first powder-plant telephone system in 1882 or 1883 and planned a clubhouse for his dynamite makers. He established himself in Philadelphia, the first member of the firm in more than seventy-five years to take up residence outside of Delaware. On his staff he had some of the ablest high explosives men in the country.

Where Howden had used sugar in "White Hercules,"

Lammot used wood pulp in his "Atlas" dynamite; for India saltpeter he substituted Peruvian or Chilean saltpeter, cheaper and more plentiful. Moreover, "Atlas" dynamite had at its back what no competitive product could hope to possess—the all-covering sales organization of the black powder triumvirate of Du Pont, Laflin & Rand, and Hazard, with its infinitude of satisfied customers and connections.

Even General du Pont became enthusiastic. "As to blasting under water," he wrote an old black powder customer, "we must frankly advise Nitro-Glycerin. We refer you to Repauno Chemical Company. Atlas powder is the best and safest made."

The Pennsylvania Railroad, long hostile to dynamite, sent a chemist to Repauno where he watched Lammot's men pitch boxes of dynamite from a tower onto rocks, set dynamite on fire and make other tests of its comparative safety, as the Nobels had done before the skeptics of Europe a decade and a half earlier. The chemist found that, with proper safeguards, dynamite could be freighted with greater safety than black powder. The railroads now swallowed the same dose that the General had found to be surprisingly sweet.

Lammot's tenacity had opened an almost limitless new field. Boldly, he and Solomon Turck stepped out to capitalize it. The Repauno company bought outright the California Powder Company's new dynamite plant at Cleveland and incorporated it as a new concern, the Hercules Powder Company. The fact that the Cleveland plant had

more than trebled its output in three years, yet was sold to
Repauno for a price that could not have exceeded $30,000,
made the transaction appear almost in the light of a gift,
with the blessings of the now converted General. The
"Hercules" plant was further enlarged. Soon it had captured 40 per cent of the explosives sales made in the Lake
Superior copper district, where the Calumet and Hecla
Mines were setting records in world production. "Hercules" dynamite also helped clear Ohio farms of the last
stumps of a once virgin wilderness. It followed the lumber-
man westward.

The "Giant" plant in upper New Jersey had been incorporated as the Atlantic Giant Powder Company. The
Repauno associates, acting as individuals, bought Atlantic
Giant stock until they owned one-third. Its reorganization
as the Atlantic Dynamite Company followed. The "Big
Three" of dynamite—Atlantic, Hercules and Repauno—
then joined in an agreement similar to that fostered earlier
by the "Big Three" of black powder. The nation was
divided in two with the states of Montana, Wyoming,
Colorado, Utah and New Mexico, with their rich gold,
silver, copper and lead mines forming a "neutral belt."
All territory to the west was assigned to the two big Pacific
Coast companies, and all eastwardly to the "Big Three."
Each company licensed the others to use its patents, an act
that ended a welter of patent litigation and opened new
vistas.

Nobel's blasting caps, protected by patents held by the
Atlantic Dynamite Company, were the key to the success-

ful use of all high explosives. The "active-dope" dynamites of the "Hercules" and "Atlas" types were pushing
the kieselguhr dynamites off the market, so that the right
to use these patents was of no small value to Atlantic.
But that was not all. Nobel, fuming in Paris over what he
regarded as the piracy of his original dynamite invention,
had himself developed a line of dynamites of active-base
materials, which were superior to all others in one highly
important field of use. Atlantic also owned patent rights
to these.

Guncotton re-entered the picture here. Thus far, it had
made little or no progress as an explosive suitable either
for war or peace. Paradoxically, however, this compound,
too dangerous for mountain-wrecking or military destruc-
tion, had been adapted by a Boston medical student, J.
Parker Maynard, for the needs of medicine and surgery.
He put it in solution with ether and alcohol and produced
the harmless chemical known as collodion. Coated over a
wound, collodion quickly dries into a thin, tough, protective film.

Nobel, like many other research men, kept some collodion on his medicine shelf. One night, in 1873, he cut a
finger. He applied collodion, watched it dry, and went to
bed. But the finger pained. The pain kept him awake. He
pondered over the new dynamite he was seeking. Long
ago he had wedded nitroglycerin and guncotton, but the
union had not turned out well. Prompted by the aching
finger, the idea of using collodion came to him. He hurried to his laboratory. Shortly after dawn, when his assist-

ant appeared to begin a new day, Nobel met him with the cry, "I've got it!" The inventor's collodion-coated finger pointed triumphantly to a jelly-like substance of yellowish tint. It was Nobel's third great contribution to mankind, the first blasting gelatin.

Blasting gelatin is almost as waterproof as rubber. It is unsurpassed for deepening rivers and harbors, and for other blasting under water. By adding to it dopes such as are used in the active-base dynamites, graded gelatin dynamites are obtained suitable to almost any sort of condition where water is present in the bore hole, as in wet mining operations.

Patented in the United States, in 1876, Nobel's gelatin explosives completed the keyboard upon which the modern blasting expert plays in the quarries, mines and engineering projects of the world. Lammot du Pont and Solomon Turck regarded the Nobel patent rights as so valuable that they agreed to pay the Atlantic company for the use of them 45 per cent of the combined profits on all dynamite sold by their companies, in addition to granting to Atlantic the rights to their own patents. They kept on buying Atlantic stock.

So wholly was Lammot occupied with the dynamite business that he resigned from the Du Pont firm. He set himself to the task of improving and making safer the methods of dynamite manufacture. Mixing of ingredients was still done by hand with rakes and shovels. With William du Pont as his aide, Lammot developed a power-driven wheel mixer, which, with minor changes, is still in general use. Next, they began work on a machine to fill and pack the dynamite cartridges, then another hazardous hand operation.

Lammot's dream was creation of a dynamite plant so thoroughly mechanized that workmen could be kept away from the zone of any dangerous operation. But his days were too few. He died March 29th, 1884, at Repauno, probably as he would have preferred to die—in action.

For years, he had been working to perfect a method to recover, in their pure form, the valuable acids left over in manufacturing nitroglycerin. Early that spring he had brought his acids recovery plan from the laboratory to the "works" stage. In a specially designed nitroglycerin or "N. G." house at Repauno, it was undergoing preliminary tests. The last charge of Friday, March 28th, had been drawn off into a lead-lined tank and left to stand overnight. This was an error in judgment that Lammot himself would not have permitted, had he been present.

Early Saturday, he went from Philadelphia to Repauno to keep an appointment with A. S. Ackerson of the St. Louis sales office of Laflin & Rand. He was in the laboratory with Ackerson and Walter N. Hill, the plant superintendent, when a workman reported trouble in the "N. G." house. Asking Ackerson to wait, Lammot and Hill hurried to the spot.

The powerful mixture in the lead-lined tank was fuming to an extent that it seemed to be boiling. Waving aside the others who were there—Harry W. Norcross, assistant plant superintendent; George Norton, foreman, and Louis

Ley, his overalled helper—Lammot and Hill entered the building. Apparently, Lammot tried to transfer the decomposing nitroglycerin into an adjoining water tank and quench or "drown" it. Anxious over the result of his experiment, he would not have taken such a risk in the regular course of manufacture.

Finally, both men retreated from the "N. G." house. They were only ten feet away when the tank exploded. Lammot and Hill were buried under a bank of earth. Norcross, Norton and Ley, about thirty feet farther away, were killed by flying timbers. Tired of waiting, Ackerson had walked from the laboratory and was approaching. His neck was broken by the shock.†

William du Pont, sobered by his sudden responsibility, became president of the Repauno and the Hercules companies. He felt lonely in Philadelphia and the offices were moved to Wilmington. Most of Lammot's stock in Repauno was purchased by the Du Pont firm. Two new young and ambitious Du Ponts joined the Brandywine powder line—Charles I., a great-grandson of the first Victor, and Alfred I., eldest son of the second Eleuthère Irénée.

Lammot's eldest son, Pierre Samuel du Pont, aged fourteen, was now the head of his family, composed of his widowed mother, four brothers and five sisters.

More cares piled upon the General. The most trusted of his agents and closest counselor outside the firm, F. L.

†This description of the accident is reconstructed from an account of it written on April 1st, 1884, by Francis G. du Pont, who with Eugene du Pont went immediately to Repauno upon learning of the explosion.

Kneeland, died in New York City. Kneeland had served as the roving ambassador of the Du Pont interests and in numerous confidential capacities within the Gunpowder Trade Association. The General had no other than himself to fill Kneeland's place. Stubbornly he refused to let even his son, the Colonel, relieve him of any of the details of the business he had shouldered so long alone.

In 1888, at the age of seventy-six, he launched one of the greatest of his enterprises. To Mooar, near Keokuk, Iowa, he sent the experienced powderman, Francis Gurney du Pont, to build the world's largest mills for the manufacture of black blasting powder. The Mooar mills were to be a model, as near perfect as Du Pont "know how" could make them—a lasting testimonial to General Henry's faith in black powder. Time was to prove the soundness of his faith. Not until 1921 did black powder decline definitely in the table of American explosives sales. Mooar, serving the soft-coal fields of the Mid-Continent, was destined to outlive even the parent mills of the Brandywine.

Much of the fire had gone from the General, however. He had allowed his clerks to force him into installing a typewriter in the plank-floored Brandywine offices. In 1889, he let the ancient oil lanterns that dimly illumined the yards by night be displaced by electric lights, though lanterns continued to serve inside the mills until after the turn of the century. The last big covered powder wagon was marked for early retirement as the General succumbed to the railroads.

One June morning, in 1889, he failed to arise from bed

as usual and had his work sent from the office to the house. On August 8th, 1889, he died. The day was his seventy-seventh birthday, the house that in which he had been born. For fifty-five years he had been a powderman. For thirty-nine years he had headed the firm. Foe of change, Henry du Pont was one of the greatest changers of his time in the explosives industry.

The New Senior Partner

A SOLEMN, almost frightened group of Du Ponts met in the bare Brandywine offices following the General's death. Seven were there. The ex-soldier, Colonel Henry A., was seated by his younger brother William, now thirty-four. The studious chemist, Eugene, was flanked by his brothers, Francis G. and Alexis Irénée II—"Doctor Alexis" by virtue of a medical degree. Standing apart, silent in the presence of their elder kinsmen, were the two youngest, Charles and Alfred.

Colonel du Pont, fifty-one, was the eldest. By seniority he was entitled to succeed his father. He was intimately acquainted with the transport problems of the company, was president and general manager of the Wilmington & Northern Railroad, was the Du Pont best posted on military and naval needs; but the Colonel was not a practical powdermaker.

Doctor Alexis knew even less of the powder line than the Colonel. Instead of practicing his profession, he had gone into the paper business at Louisville, Kentucky, after receiving his degree at the University of Pennsylvania, and had returned to "the Brandywine" only four years ago. Now forty-six, the doctor viewed with only an academic interest the grime and danger of "the works."

The mantle of the late General fell, therefore, to the reserved, unostentatious Eugene. Since his graduation from the University of Pennsylvania, his life had been spent in the Brandywine mills—as assistant to Lammot, head of the laboratory, and lately general works manager. Two patents on a new gunpowder, called brown prismatic or cocoa powder, were largely the result of his efforts. This, a variation of black powder, was given its color and a fire superiority by an improved type of charcoal. He was forty-nine.

Eugene spoke little. His sober bearing gave the impression of austerity, though he was a kindly man. Men of the works spoke of his coolness in emergency. In 1882, he had prevented what might have been a serious explosion at no small risk to himself. A fire in the laboratory of the saltpeter refinery was sending showers of sparks over a black powder graining mill just below. Gathering a group of employees, Eugene stood calmly in the mill doorway until the last bag of powder was thrown into the race and the mill thoroughly saturated with water. The men, not the silent chemist, later revealed what had been done.

No other was as well suited as Eugene to assume the leadership. That, the assembled Du Ponts agreed.

Throughout his long service, the General, and he alone, had charted the company's course. As senior partner, and ex officio head of the family, he had been bound to accept nobody's advice, and, by family tradition, nobody's advice had been offered him. In person he had opened all company mail, in person he had answered it. He had written all checks, negotiated all contracts, had been the judge in all controversies, alone had known all details of the firm's affairs. After the day's work, members of the partnership had dutifully met in his small, white-walled office —the black rim of powder dust on the wall back of their chairs was evidence of the regularity of their meetings. However, the partners usually had come to give information, rarely to get it.

Eugene du Pont faced a difficult task in taking over the General's vacated post. Beyond his own limited sector, he was inexperienced in the business world—and the complexities of management were mounting. Four clerks and a stenographer were the only assistants in the office, which left him with responsibility even for the trivial details of administration.

Graphic evidence of the Spartan-like regime that had preceded was the firm's letter paper. It was unprinted. The founder, Eleuthère Irénée du Pont, had used the same kind of plain, nondescript white sheets on which, below the signature of the senior partner, was written, or later typed, the name, E. I. du Pont de Nemours & Company. Some letterheads of Du Pont sales agents, on the contrary, were flamboyant examples of the printer's art. Eugene noted the prominence frequently given to the agent's name, the much smaller type accorded the Du Pont name. Changes were in order, he decided.

A new office building, more representative of the company, was authorized to replace the little four-room structure that had served more than fifty years. A telephone line was installed; an additional clerk and stenographer engaged. In 1897, Eugene yielded to his partners' insistence and brought his son, Eugene, Jr., recent graduate of Harvard University, into the office to assist him in administrative details. This was the first break from the Du Pont tradition that young Du Ponts should enter the business at its source, the powder line.

Eugene introduced other reforms. Under the General, a small number of carpenters and masons had been maintained on the payroll. Part of their time was spent on work connected with the commonly owned farms. Miles of evenly laid "dry" stone walls that still bound fields in the vicinity of "the Brandywine" were one evidence of their labors, which the new senior partner decided was not powder-making. He discharged the men, directing that thereafter all such work should be let out to contract. A few of the carpenters protested and resorted to violence.

On the night after Christmas, 1889, a new barn near Francis G. du Pont's home was fired by incendiaries. A second barn blazed a week later, two more in following months. A powder works is not a comfortable place when angry men are carrying the torch by night. The worst was expected—and came. On October 7th, twelve persons were killed and twenty injured in seven explosions within eight seconds that wiped out half of the Upper Yard and 200 tons of gunpowder.

Investigation cleared the barn-burners of any part in the disaster at the powder yards. An overheated soldering-iron, being used to seal tin powder-boxes, had started the explosions. This operation was specified by the United States Government. A solder of low melting point was used and the operation was considered safe in the hands of two old and trusted men. But it was a windy day and perhaps a spark from the necessarily near-by soldering-furnace was responsible for the tragedy that followed.

However, several of the carpenters were implicated in the incendiary fires. They were prosecuted, convicted, and jailed. The company's first labor trouble in almost ninety years was ended.

Of the accident:

"Concerning the losses and needs of all our hands," Eugene wrote, "we propose to them and their families and to all who have suffered, to bring out of chaos an orderly state of affairs, to restore everything except life to all, to nourish, protect and guide all and to do everything possible for man to do."

This was the traditional obligation.

Still Eugene's troubles were not over. In 1892, William du Pont, Lammot's old ally and successor as the president of the Repauno company and its affiliates, resigned and retired from active service with the company. Looking about, Eugene saw no Du Pont available or sufficiently schooled to head the dynamite enterprise a Du Pont had founded.

Laflin & Rand, the co-owners, were consulted. They suggested J. Amory Haskell, a young coal-mining executive who was scarcely thirty-one years old. His record was brilliant. As general manager of the Rochester & Pittsburg Coal & Iron Company during a period of financial panic and troublous labor conditions, he had produced both harmony and profits, and, otherwise, demonstrated marked qualities of leadership. A brief period as vice-president and acting president resulted in young Haskell's election to the Repauno presidency without a dissenting voice.

The Repauno company's secretary was Hamilton M. Barksdale, another thirty-one-year-old. Mutual friendship and respect, destined to continue throughout their lives, sprang up upon his first meeting with Amory Haskell. Shortly, in all but the title, Barksdale was his new chief's general manager. Neither of them was a chemist or powderman. Haskell was not a college graduate. However, both of them knew men, and business.

Over six feet tall, broad-shouldered, Haskell spoke the language of railroad and mine, great major users of explosives. He knew their needs intimately. The keen competition of the soft-coal industry, wherein profits were the reward only of the most vigilant attention to costs, had sharpened both his ingenuity and perspicacity. He was a born organizer. Barksdale was a civil engineer, graduate of the University of Virginia, and he had helped construct the Baltimore & Ohio railroad's new line between Baltimore and Philadelphia. He, too, was acquainted with explosives as a customer. Men liked him instinctively.

Throughout the history of the Du Pont Company, it has been customary to trust men unqualifiedly once they have been found worthy of trust. This was what happened when Eugene du Pont, steeped in the old firm's traditions, accepted Haskell and Barksdale. They were given untrammeled authority. They found themselves in a field ripe for tilling. The high explosives business was in its infancy. Methods were simple. Producers devoted themselves to filling orders, while actual sales were made through jobbers or in special instances "consignment agents." The manufacturer had little or no knowledge of how or where his product was to be used, and except to the extent of improving quality, had only a casual interest in his ultimate customers' problems. This was the "comfortable" era, when not only explosives makers but most manufacturers rarely looked beyond their own shops.

It was a day of "trade secrets," rule-of-thumb, and the "lone inventor" usually struggling without funds. Organized research was not even a name. Chemistry, judged by modern standards, was in an elementary stage. Organic chemistry was regarded in the light of an occult art and indulged in chiefly for its academic interest. The organizer, Amory Haskell, had plenty to organize.

By 1895, he had importantly reshaped the structure of the high explosives industry. Largely under his direction, that year, the Eastern Dynamite Company was formed with a capital stock of $2,000,000. Of this, $1,400,000 was exchanged for the capital stock of the Repauno and Hercules companies, and $600,000 for the assets of the Atlantic Dynamite Company, holder of Nobel's American patents. Atlantic was then re-incorporated as a wholly owned subsidiary of the new Eastern Dynamite Company, which became merely a holding company. Also that year, the aggressive young chief of Repauno succeeded the veteran Solomon Turck as president of Laflin & Rand.

This reorganization in no way disturbed the primary control of the "Big Three" of dynamite by the "Big Three" of black powder, namely Du Pont, Hazard, and Laflin & Rand.

With Barksdale as his liaison officer, Haskell quickly introduced any improvements made by one company to the others of the combine; and these two were geniuses in devising improvements. For the first time in the industry investigations of customers' needs were carried straight to the customer himself. The "agency" system was abolished, company branch sales offices were established in important cities, and company salesmen and demonstrators were put on the road to make direct contacts with dealers and big industrial users of dynamite. Haskell expounded the doctrine of "low prices, large volume," which was to find its ultimate expression years later in what was to become known popularly as "mass production."

Following Lammot du Pont's ideas, machinery was developed to displace slow hand operations, which not only increased production but also reduced hazards and costs. Responsible for numerous improvements of this character was the painstaking, blunt-speaking plant superintendent, Oscar R. Jackson, whom Haskell felt was the ablest chemist in the industry. Son of Dr. Charles T. Jackson, who first discovered and successfully used anæsthetic ether, he was named for King Oscar I of Sweden, because of decorations the king had bestowed on his father. Trained at Harvard and in Germany under the celebrated chemist, Adolph Bayer, when Oscar Jackson spoke, he knew; when he did not know, he found out. As Repauno's superintendent, he was czar in his domain.

More technically trained men were brought into the business, young men with ideas and eyes open to the expanding field of the usefulness of dynamite. Among these were Harry M. Pierce, recruited from the Wilmington shops of Pusey & Jones, and Harry G. Haskell, younger brother of Amory, graduate of the Columbia University School of Mines. So fast was the technical pace at Repauno that, after three years, young Pierce asked for and received a leave of absence to attend the University of Pennsylvania's Engineering School. Sound technical-school training was beginning to be regarded as indispensable to getting ahead in the industry, which was a long advance from the day when it was believed that powder-making could be learned only on the powder line.

Organization was yet to enter the administrative offices, however. "The business," Harry G. Haskell wrote later, "was conducted by a few people who around-the-table read all the mail and attended to everything in somewhat haphazard fashion."

Barksdale suggested they introduce system in the office as well as the plants. "Suppose you take the operating end," he told the younger Haskell, "and I'll handle sales, or the other way around, if you prefer."

Forthwith, Harry G. Haskell began organizing the High Explosives Operating Department. He did not know it, but he was taking his first step toward becoming the general manager of all of Du Pont's high explosives interests, of which the Repauno plant was to be the nucleus and model.

In this intimate, closely cooperating management group they discussed another innovation—the matter of rewarding able men beyond their salaries through special grants or "bonuses." The elder Haskell felt strongly on the subject. It was his thought that ownership should be distributed as widely as possible over the company's management to heighten self-interest in the enterprise. No plan was formalized, but, during this final decade of the century, one of the first experiments with a bonus system was carried on at Repauno.

A sign of the industry's rapid technical progress was the fact that less than ten years after Lammot du Pont had remodeled it, the Cleveland plant of the Hercules company was condemned as obsolete and scrapped. A new plant was erected near Ashburn, Missouri, in time to supply a major share of the more than 15,000,000 pounds of dynamite consumed in blasting the Chicago Drainage Canal. Dynamite consumed on this single project was greater than one-half of the total output of black powder for all purposes in the nation in 1860. The amount of "Atlas" and "Hercules" dynamites supplied was greater than the total amount of gunpowder the Du Pont Company had sold the United States during the four years of the Civil War!

In a significant respect this hurrying, daring, youth-driven dynamite industry was different from the old black powder business. Dynamite is distinctly an industrial tool. It may be used incidentally in war as a demolition agent, but it cannot be fired successfully in guns. Amory Haskell's young men included not a military expert. His customers were engineers, mine operators, quarrymen, farmers, dredgers, builders in every conceivable field, fields expanding so fast that the industry only with difficulty could keep its production in pace with demands.

The needs of railroading and mining were soon to be augmented by an unparalleled expansion in highway building. Rise of the automobile and the general use of steel as the framework for large buildings were to send the steel industry's consumption of dynamite to heights not even remotely dreamed.

When Haskell took command of the Repauno interests, the nation's total output of dynamite was approximately 30,000,000 pounds. Within a decade, American dynamite production exceeded 130,000,000 pounds. In 1914 it was 262,000,000 pounds, in 1924 almost 345,000,000 pounds. In thirty years Haskell saw the market with which he began multiplied more than eleven times. And it was an industrial market exclusively! He saw dynamite-making become, through rigid precautions and technical advance, as free from accidents as most other kinds of chemical manufacture.

★ CHAPTER ★
V

Cellulose Marches

MEANWHILE, chemistry was coming to the fore. Schoenbein's guncotton had started a march that was to lead to startling upheavals, reshape industries, change the work and habits of millions, alter the trade of nations, reorder the science of war, and, incidentally, rebuild the house of Du Pont.

The proper chemical term for guncotton is cellulose nitrate. The facts that cotton, chemically, is almost entirely cellulose, that cellulose makes up the fibrous or cellular structure of most plants including trees, and that it is inexhaustibly abundant in the world, made Schoenbein's discovery of interest in fields far removed from explosives.

Following Maynard's adaptation of cellulose nitrate for use in treating minor wounds, millions of bottles of collodion were sold. One such bottle, as we have seen, got into the medicine cabinet of Alfred Nobel and out of it came blasting gelatin. Prior to this, in 1863, if you had poked about a printing-shop in Albany, New York, where was employed one John Wesley Hyatt, you would have come upon still another bottle of collodion. Corked within this bottle was a second genie more powerful than the one Nobel raised.

John Hyatt was twenty-six, printer son of a country

blacksmith. Most of his education had been acquired in setting type. In 1863, an advertisement of the Phelan & Collander Company offering $10,000 for a new material with which to make billiard balls started young Hyatt on a search for artificial ivory, long a dream of chemists. James Brown, a fellow printer, joined him.

Working in spare time in the kitchen of Mrs. Mac-Tavish's boarding-house, Hyatt first tried to press billiard balls out of pulverized wood. His only success was in making such an odor that Mrs. MacTavish forbade further use of her kitchen. Hyatt rented a shed and continued his experiments there, to the alarm of his neighbors. A local chemist discovered that the two printers were trying, as he reported it, to press billiard balls out of guncotton.

Hyatt and Brown moved to a more isolated shed and this time bolted the door. They were not working with guncotton at all, but with collodion, or pyroxylin as it is better known today. That bottle on the printing-shop shelf had given Hyatt a clue. One evening, after a day of typesetting, he had found the bottle overturned, its contents spilled and hardened. For the first time he was attracted by the ivory-like nature of the dried material, which he knew to be a weakly nitrated cotton dissolved in ether and alcohol.

Interested, he studied literature dealing with cellulose nitrate. He found that, in the form of pyroxylin, the compound is inflammable, but that it is not explosive. Moreover, he learned that a plastic material resembling ivory had already been produced by Parkes, an English chemist, by combining pyroxylin and camphor. This was the fundamental discovery of the wholly new material Hyatt envisioned.

However, the English "Parkesine," as it was called, was not commercially usable because of its high cost. Parkes employed a comparatively large quantity of solvent, the recovery of which still awaited development of a feasible method. Pondering over this detail, Hyatt conceived the idea of using only a small amount of solvent and supplementing it with heat and pressure.

Behind a flimsy wooden shield set up against the possibility of accident, the two printers put the idea to test in a rude press they had built. The result was "Celluloid," the first of the cellulose plastics sufficiently low in cost to be suitable for large-scale manufacture. Thus, humbly, was born the plastics industry, modern marvel of chemical fecundity, which has given us literally thousands of new and useful articles. Hyatt's discovery was patented in 1870. By then, Schoenbein was dead. But cellulose nitrate in its several forms kept marching.

Hilaire de Chardonnet, Count of France, had also noticed the satin-like smoothness and flexibility of dried collodion or pyroxylin. As a youth, he had studied under Pasteur, in L'École Polytechnique. He knew of Pasteur's work on a silkworm disease that at one time threatened to destroy the French silk industry. If man, he pondered, could make a silk-like fiber independently of the silkworm, science would no longer need to worry itself over the diseases of worms. The thought, even then, was more than two centuries old. Scientific literature was filled with the recorded failures of others to produce a synthetic textile fiber. But like Parkes and Hyatt, the Count de Chardonnet had an original idea.

He constructed a spinning machine, the heart of which was a metal disc, or "spinneret," through which had been bored a number of tiny holes. Into the machine he poured pyroxylin. The syrupy solution was forced through the tiny holes until it squirted out in fine sprays, into a stream of hot air. The air evaporated the solvents—ether and alcohol—and left the cellulose nitrate in the form of light, fluffy, lustrous filaments. It was the first rayon made, although that word was not to be coined until 1924.

De Chardonnet spent five years improving his machine and developing a way to reduce the nitrate content of his fiber to render it less inflammable. The public announcement of his invention was not made until the International Exposition at Paris in 1889, when it created a sensation. Oddly, the first practical use of the new fiber was not in textiles, but in incandescent lamps, where threads of it were used for carbon filaments—it was still much too crude and costly for textile purposes. However, the seeds of a textile revolution had been sown. And cellulose nitrate was still marching!

It marched, now, right up to the Du Ponts' front door. Since the first Alfred du Pont had tested guncotton, forty-odd years earlier, and pronounced it too violent an explosive for use in rifles and cannon, the discovery of the pyroxylin-type plastics had changed the picture materially. If Schoenbein's touchy agent of death could be tamed to a docility that permitted it to be made safely into babies' rattles and white collars, its violence might also be tempered in the intermediate degrees by application of the same principle, namely, use of less solvent and then employing heat and pressure to increase the solvent action and the plasticity of the material. Experimenters the world over hastened to adapt that discovery to the conversion of cellulose nitrate into a base for an improved gunpowder. Smokeless powder was the result, the base of which is pyro-nitrocellulose, a type of cellulose nitrate that is intermediate to the highly explosive guncotton and the non-explosive pyroxylin.

Superiority of smokeless powder over black powder in firearms is as marked as the superiority of electricity over kerosene in illumination. Good smokeless powder becomes wholly gaseous in combustion and burns fully, cleanly, and at almost any desired speed. Its invention and perfection, which grew out of the work of many experimenters and may be attributed to no one person, removed at a stroke the chief obstacles—smoke and dirt—that for years had been blocking the successful use of long-range, rapid-fire, rifled guns.

Vieille, a Frenchman, in 1886 compounded the first military smokeless powder of the modern type. His *Poudre B* was a straight pyro-nitrocellulose variant of Hyatt's pyroxylin plastic, with certain chemicals added to promote combustion and stabilize the mixture. *Poudre B* was known as a single-base powder. About two years later Nobel invented a compound of pyro-nitrocellulose and nitroglycerin, or a double-base powder trademarked "Ballistite." The controversy still rages over the respective merits of these two classes of smokeless powder.

Russia and Belgium each adopted a modification of the French *Poudre B* or single-base powder. England, Germany and Italy, on the other hand, developed double-base powders on the principle of Nobel's "Ballistite." Italy and England borrowed the idea of the spaghetti-making machine and formed their powder into strings or cords, from which was derived the name of the famous British "Cordite." The Italians called their similar powder "Filite."

Amid the controversy, the United States Navy experimented with single-base powders, while the Army experimented with double-base powders. Each guarded its work jealously, but on one point they were agreed. America was without a satisfactory smokeless powder of any kind.

In 1889, the Army's Chief of Ordnance asked the Du Pont Company to send a competent powderman to Europe to try to purchase the secret formula of the French *Poudre B* and also investigate the European powder situation generally. Twenty-five-year-old Alfred I. du Pont, the eldest son of the eldest son of the first Alfred, was selected for the mission. In France, he failed to get any data on Vieille's closely guarded formula. The English likewise were unwilling to disclose any of their secrets. However, the Belgian firm of Coopal & Company were quite eager to sell the secret formula for their single-base powder, which they claimed was the best in Europe, although they expected the buyer to accept that claim on faith.

Eugene du Pont questioned Coopal & Company's high price, shied from what seemed to him a pig-in-a-poke proposal, but the worried Army chiefs insisted and Du Pont bought the Belgian formula. A few months later Eugene wrote off the investment as a total loss. In tests, the Belgian powder was surpassed by a new double-base powder invented by Hudson Maxim, brother of the inventor of the Maxim machine gun. However, Maxim's powder was not entirely satisfactory either.

Eugene decided that if the Du Ponts made a smokeless powder at all, it would have to be one of their own development. The veteran Francis G., just returned from Iowa where he had started the wheels of the Mooar blasting-powder mills, was assigned to the problem. He set up a laboratory and went to work, assisted by Pierre Samuel du Pont, serious-minded, mildly spoken eldest son of Lammot. The ink was hardly dry on Pierre's M. I. T. diploma and he was not yet old enough to vote.

That year, 1890, a large tract of flat open land was purchased at Carney's Point, New Jersey, across the Delaware River from Wilmington. The next spring a more elaborate laboratory was built there, and, at the request of the Navy, a plant was begun to supply guncotton for naval mines and torpedoes. In June, 1892, the first guncotton was produced. By the year's end, 100,000 pounds of it had been shipped to the Navy.

But guncotton was not smokeless powder. Not until 1893 were Francis G. and young Pierre du Pont able to report they had compounded a smokeless powder that would perform satisfactorily—not in rifles, pistols and cannon—but in hunters' shotguns. In 1894, this first Du Pont smokeless sporting powder was put into commercial production, in time for the rabbit season.

The Du Pont Company was now interested in cellulose nitrate on three scores: as a base for blasting gelatin and gelatin dynamite, as guncotton for torpedoes and mines, and as a propellent base for shotgun shells. In the field of commercial explosives the company and its connections were pre-eminent; but in the development and introduction to America of new munitions of war, they were hardly of secondary rank.

It is not difficult to understand the firm's reluctance to plunge into an intensive program of military powder development. War seemed remote. For almost a century the market for industrial explosives had been increasing at a breath-taking pace. That pace was still accelerating. Great engineering projects underway or in prospect, continued industrial expansion, the national awakening to the need for highway improvement—these and other works of a people at peace were making demands upon the explosives industry that it was hard-pressed to meet. The normal military need had become a small tail on a big dog, and there was every reason to believe it would continue so indefinitely, except in the case of a most extraordinary emergency.

In consequence, the main burden of developing a military smokeless powder in America during the Nineties fell upon the Army and Navy ordnance staffs and independent inventors. The Du Ponts and other large powdermakers cooperated, but their research efforts were directed more toward improving sporting powders, with the military a secondary consideration.

To Eugene du Pont, however, military powder-making was no longer a business but a duty, like jury service. Carney's Point became a training school for Du Pont youngsters who wanted to serve—Pierre; Eugene's own son, Alexis I. 3rd; and two sons of Francis G., Francis I. and A. Felix.

The Spanish-American War, which started in April, 1898, precipitated a crisis. The Navy's magazines were empty. Ships lacked enough cannon powder for a two hours' major engagement. The war caught the nation midway in the experimental phases of a transition from an old type of gunpowder to a new, still incompletely developed. It took six months to dry smokeless cannon powder once it was made, and thereafter its keeping qualities were uncertain. It was decided, therefore, to fight the war with a brown prismatic powder, the basic patents for which were controlled by the Vereinigte Koeln-Rottweiler Pulverfabriken, of Germany.

About ten years previously, again over Eugene du Pont's objection but at Washington's behest, the Du Pont Company had bought the secret formula for this German powder upon the Government's pledge to pay the royalty of

3 cents per pound. The California Powder Company was also licensed to produce the powder, so that two plants existed for its fabrication. The Du Pont capacity was only 3,000 pounds per day. Suddenly, the company was requested to increase this capacity more than eight times—to furnish 5,000,000 pounds of brown prismatic powder with the utmost dispatch possible. The price fixed was 29 cents, plus the German royalty.

For a fourth time in the century, the Brandywine Works went on a war basis for the Government. Outside shops, called upon to help build additional machinery, worked night and day. Commercial powder production was suspended. New buildings had to be erected, powder containers prepared. Faster presses, the invention of Alfred I. du Pont, were hurriedly made and installed.

Sixty days after the declaration of war, the Brandywine mills attained a brown powder output of 25,000 pounds every twenty-four hours. Within four months they delivered 2,200,000 pounds of powder to the United States, chiefly for the Navy, which bore the brunt of fighting. Then the war ended. Still under contract, with facilities standing ready to make it, were orders for more than $800,000 worth of an outmoded military powder that nobody wanted.

On August 15th, 1898, Rear-Admiral Charles O'Neil, Chief of the Navy's Ordnance Bureau, wrote the firm asking permission to cancel his last order for 1,000,000 pounds of brown powder, and substitute for it a similar amount of smokeless powder "at a price which will compensate you for the expense to which you have been put in increasing your output of Brown Prismatic."

Eugene du Pont replied:

"For the last four months we have devoted the whole output of our mills to the Bureau and Ordnance Department. Our customers who have been with us for years have been scantily supplied. We have jeopardized our business to a great extent. As to compensation, we fully believe that our business will be benefited to a much greater extent by turning our mills onto the regular manufacture than it would to continue to make Brown Prismatic powder. We, therefore, think that no compensation is required for the money expended in increasing the output of Brown Prismatic powder."

Whereupon Eugene canceled all brown powder contracts and swung the Brandywine mills back to "useful, orderly business." The war-day plants were scrapped. Future wars were destined to be fought with nitrocellulose—millions of tons of it. Oddly, Eugene du Pont never envisioned this. As late as 1900, he wrote, with unmartial naïveté, that the "smokeless military powder business has seen its maximum."

Sometime before this, in the spring of 1899, Lammot's son, Pierre, now twenty-nine, had resigned his post at Carney's Point and left Delaware for distant fields. He saw little future for an ambitious young man in the Du Pont Company as then constituted. Pierre's going was a danger sign, but Eugene did not see it.

As senior partner, he had unswervingly continued the expansive policies of his predecessor. Since 1889, company after company had been acquired outright or brought under control, directly or through the Eastern Dynamite Company, or in cooperation with Laflin & Rand. The details of administering the extensive Du Pont interests had mounted to a point where no one man could handle them alone without staggering. But alone Eugene did handle them, as dictated by tradition. He was the head of the firm!

Finally Colonel Henry du Pont called a halt. In the settlement of his father's estate, he had been appalled by the labor the General had been through in his latter years. He urged that the partnership, formed in 1837, be changed to a corporation and the administrative responsibility be more equitably distributed. Eugene was won over to the idea. On October 23rd, 1899, E. I. du Pont de Nemours & Company was incorporated in Delaware and the old partnership passed into history.

Actually, the change was in form only. Eugene, as president, went on as before. He was too conscientious to delegate responsibility that he felt was his alone. His brothers, Doctor Alexis and Francis G., were named vice-presidents, along with Colonel Henry A. The first Victor's great-grandson, Charles I. du Pont, was designated secretary and treasurer, the younger Alfred I. du Pont, a director without title. But no others were admitted to ownership, the executive staff was still limited to the Du Pont family; only the elder Du Ponts sat together in conference. One man still ruled. Personally, he continued to conduct all matters involving policy and carried on all correspondence about them in long hand with a copy press.

Two years and two months later—January, 1902—Eugene du Pont was stricken with pneumonia. He died within a week.

Colonel du Pont was asked to take the presidency. He declined. In late years his outside interests—the railroad he headed, concern over the ex-soldiers and their problems, and reform of Delaware politics, then in turmoil—had gradually outweighed his interest in the business. The Colonel was on his way to the United States Senate (1906-17), and, afterward, to the leisurely existence of a country gentleman on his estate, "Winterthur." He was to live to be eighty-eight.

Both Doctor Alexis and Francis G. du Pont were ill when Eugene died. Francis was despondent. Less than three years of life remained for them—November, 1904, was to see them go almost together. In poorest health of all was Charles I., still in his thirties; he was to die before the year ended. That left only Alfred I. Despite a robust constitution and recognized ability as a powderman, he was regarded as too young, erratic, and inexperienced in business to fill the presidency. He himself recognized that his years in the "powder yards" had not fitted him to take charge of the administrative duties of the company.

The three elder Du Ponts faced each other that winter day of 1902, dismay and futility in their eyes. They had given their youth, strength, very souls to the company. Yet in its moment of great need they felt impotent.

The company seemed without a Du Pont to become its leader. It was suggested that the presidency be offered to

Hamilton M. Barksdale. He had married Charles I. du Pont's sister, Ethel, and was therefore a member of the family by marriage. Since 1887, he had been importantly identified with Du Pont affairs. Now forty years old, he was known throughout the industry as an outstandingly able executive. Only to one other man in a century had a similar proposal been made, a rare mark of confidence, but Barksdale felt that a "man of the name" should be elected to the post.

Wearily, Francis G. du Pont gave words to the only other action he saw open—sale of their heritage to Laflin & Rand, the one company still active in the explosives industry in America whose beginning antedated Du Pont's own. It was a sad day on "the Brandywine," that winter day of 1902.

*

BOOK THREE

*

A New Century

Three Cousins

Sell the company? Alfred I. du Pont rebelled at the thought. He felt the business his both by heritage and hard-earned right. He would fight for it!

"I realized," he related later, "that if I proposed any plan to the older members of the directors it would undoubtedly be antagonized, and my only hope of success lay in apparently falling in with the plan suggested by Mr. Francis G. du Pont. . . . Perhaps a week later a meeting of the stockholders—which were analogous to the directors—was called for the purpose of taking action on this proposed disposal of the company's assets. . . . A resolution was offered to the effect that the assets should be sold to the Laflin & Rand company, and Mr. H. M. Barksdale was appointed as the agent to negotiate the sale. I suggested an amendment to the resolution, in effect that the business should be sold to the highest bidder. There was no opposition to this amendment, and the resolution passed the stockholders' meeting, receiving their unanimous approval in that form.

"After the meeting had adjourned, I arose and stated to the directors that I would buy the business. As you can imagine, it caused some consternation and a good deal of surprise. Mr. Francis G. du Pont made a remark to the effect that I could not have it. I asked him, 'Why not?' I pointed out to him that the business was mine by all rights of heritage, that it was my birthright; I told him that I would pay as much for the business as anybody else, and, furthermore, I proposed to have it. I told him that I would require one week in which to perfect the necessary arrangements looking toward the purchase of the business, and asked for that length of time."

Colonel Henry A. du Pont looked at Alfred, when this stand was made, with the awakening interest of one fighter in another. He arose.

"Gentlemen, I think I understand Alfred's sentiment in desiring to purchase the business. I wish to say that it has my hearty approval. I shall insist that he be given the first opportunity to acquire the property."

The head of the family spoke.

Several days later, Alfred I. met with two of his first cousins in Wilmington. Summoned by long-distance telephone from widely separated points, these two eldest sons of his father's brothers had dropped everything to answer his appeal for help.

The one, Thomas Coleman du Pont, had just passed thirty-eight and was Alfred's senior by only five months. Six feet four inches tall, weighing 250 pounds, as forceful mentally as he was vigorous physically, Coleman quickly dominated the little conference. The other cousin was Pierre Samuel, who, as we have seen, had left the company three years earlier. Pierre was now thirty-two.

It was a well-balanced trio. Alfred I. was thoroughly schooled in the manufacture of black powder and the peculiar requirements of the explosives industry. Some of the most important advances made by the company in the mechanics of powder production had been inspired by him. He was liked by the men in the mills.

Coleman knew nothing of powder-making, but he had a genius for organization and management that was seasoned by an experience more diversified than that of any

Du Pont in a century. His father, Antoine Bidermann du Pont, was the youngest son of Alfred Victor, the second Du Pont to head the business. As a young man, he had settled in Louisville, Kentucky. There he had married Ellen Susan Coleman and had become interested in paper manufacture, coal mining and finally in street railways. There, on December 11th, 1863, his eldest son, Thomas Coleman, had been born with all of his father's restlessness and love of adventure.

As a student at Urbana College in Ohio, Coleman had captained the football and baseball teams, had been stroke oar in the crew and a sprinter on the track team. He could box, wrestle, swim, ride and shoot. He was famous for his practical jokes. Later, at Massachusetts Institute of Technology, he had qualified as a mining engineer and became friendly with his cousin Alfred, then also a student at "M. I. T."

Eight years of work in the coal mines at Central City, Kentucky, had followed. From mule driver, blacksmith and miner, Coleman rose to superintendent of the Central Coal & Iron Company. His energy and vision helped lift that company from mediocrity to importance. Next, he had turned to steel. Arthur J. Moxham and Tom L. Johnson, later famous as a reform mayor of Cleveland, made him general manager of the Johnson Company of Johnstown, Pennsylvania, from which later was to spring the Lorain Steel Company, now part of the United States Steel Corporation.

Moxham was a shrewd promoter. Watching him operate, learning quickly, Coleman soon decided to launch out independently, which he did by buying the Johnstown street railway, a primitive affair in need of better management. Shortly, the railway was earning profits, whereupon he resigned from the Johnson Company, organized his own forces and entered into the electric street railway business on a national scale.

When Alfred I.'s urgent telephone call for help reached him, Coleman was not wholly unprepared for it. Several years earlier he had taken a house in Wilmington, foreseeing that the day might come when "the Brandywine" could use his services.

"Count me in with you, if I can get Pierre to handle the finances," he advised Alfred, through the wheezy instrument fastened against the wall. "Don't worry," he added, "I'll get Pierre."

Then Coleman put through a long-distance call to Pierre in Lorain, Ohio.

Pierre was even-tempered, serious, consistent. He was a quiet, precise man who never stepped until he knew the exact nature of the ground ahead. Since that March day, in 1884, when his father was killed at Repauno, he had faithfully discharged the responsibility that devolved upon him as the eldest son and family head. His four younger brothers and five sisters looked to him for guidance. To all of them he was known affectionately as "Daddy," and so consulted even in later years.

Upon leaving the Du Pont organization in 1899, Pierre had been invited by Tom L. Johnson and Coleman to enlist with the Moxham-Johnson-Du Pont interests. The prospects of large developments in the affairs of these men appealed to Pierre and he accepted the presidency of the Johnson Company. The latter had disposed of its steel holdings and was in process of liquidating that portion of the business and expanding into other fields, including street railways, when Coleman put through the call to Lorain, then Pierre's headquarters.

Pierre, too, had foreseen the emergency on "the Brandywine," although he had not expected it so soon. In him was a deep sentiment, a loyalty to family and the business of his father's that submerged personal interests and memories of past disappointments.

"I'll go," he said quietly into the telephone immediately the situation back home was made clear.

So, two days later, Coleman and Pierre sat down with Alfred I. in Wilmington. The three cousins had little concept of the value of the company they proposed to buy. They knew nothing of its numerous stockholdings and connections. They had never seen the company's books, its financial statements nor its customer lists. However, they had complete faith in the ability and integrity of the Du Pont elders and in the company's soundness. The situation was expressed by Pierre in a letter to Irénée du Pont, a younger brother.

"We have not the slightest idea what we are buying," Pierre wrote, "but we are probably not at a disadvantage, as I think the old company also has a very slim idea of the property they possess."

The one solution the cousins could see open, in the limited time for which Alfred had asked—one week—was to induce the elder Du Ponts to accept a small cash payment and a large block of promissory notes, to be secured by the stock of a new company to be formed. Coleman was delegated to go to Colonel Henry du Pont with this proposition. He did immediately.

Not only was their proposal accepted, but the elder Du Ponts waived cash payment entirely in return for a one-fourth interest in the proposed new company, and named a tentative price of $12,000,000 for the property. Even a cursory examination of the books by Pierre revealed that figure to be generously low. Appearance on the scene of the competent Coleman and of the serious, careful son of Lammot had swept aside the elder Du Ponts' last doubt of the ability of the next generation of Du Ponts to carry on with success and credit.

On March 1st, 1902, not a month later, Pierre stopped in at the company's office to discuss further the details of the proposed sale and purchase. Francis G. du Pont was there, but the mail, piled high on Eugene's old desk, had not yet received attention. Coleman was away, hurriedly settling his own affairs. As yet, the negotiations had not, legally, passed the preliminary stage. Only the tentative price had been named. No appraisal had been made. Not a paper had been drawn. Nothing had been signed.

Francis G. du Pont arose, drew back the president's chair and motioned to Pierre to be seated. "You may as well begin now, my boy," he said. "From today, the com-

pany is under the management of yourself, Coleman and
Alfred. We are turning over everything to you, with full
confidence."

To the astonished Pierre he indicated what matters
demanded action. He made a simple announcement to
the wondering office force. Then the old powderman
pulled on his overcoat, shook Pierre's hand, and left. The
mail still waited.

Pierre summoned Alfred from the powder-mills. They
wired Coleman of the totally unexpected turn of events,
and he took the next train back to Delaware. Thus, weeks
earlier than they had anticipated, the cousins were actually
running the company, and not a dollar had been paid
down by them.

As finally consummated, purchase of the assets was
made at the $12,000,000 figure originally named and paid
in 4 per cent notes. In addition, the cousins turned over
to the former owners, as a bonus, $3,000,000 in the stock
of the new corporation, E. I. du Pont de Nemours Company of Delaware.* Coleman became president; Alfred,
vice-president; Pierre, treasurer; and Charles I. du Pont,
secretary.†

Then Pierre dug into the books to find out what they
had bought. For several months he worked with growing
amazement. For the first time, perhaps, since the Civil

*This company assumed the full name E. I. du Pont de Nemours
& Company on the dissolution of the original corporation of 1899, following the sale of its properties.

†Charles I. du Pont died in November, 1902. He was succeeded as
secretary by Alexis I. du Pont, 2nd.

War, a complete, orderly record was compiled of the assets that had been built up by General Henry and Eugene du Pont and paid for out of the savings they and other members of the family had accumulated. The company itself owned outright and operated five plants—one at Sycamore, Tennessee, one at Mooar, Iowa, one at Wapwallopen, Pennsylvania, one at Carney's Point, New Jersey, and the parent plant on "the Brandywine." These plants manufactured and sold 36 per cent of the black blasting powder used in the United States. The company also supplied, through the Carney's Point operation, a major share of the newly developed smokeless powder and guncotton then being used in small quantities by the Army and Navy. However, the five plants and all attendant property represented only about 40 per cent of the company's total assets. Sixty per cent was in the stock of other powder companies.

The Hazard Powder Company was owned outright. But more than two-thirds of Hazard's assets were also represented by outside investments. Unscrambling the puzzle, Pierre found that, together, the Du Pont and Hazard stockholdings comprised minority interests in sixteen larger companies and majority interests in two, apart from their own concerns. Furthermore, in thirteen of the eighteen larger companies in which Du Pont and Hazard were interested, Laflin & Rand was also a substantial minority stockholder; it controlled two other companies independently, and had important interests in still two more, which made a total of seventeen companies in which

Laflin & Rand had investments. That was to say, the combined holdings of the three companies—Du Pont, Hazard, and Laflin & Rand, together with the holdings of their controlled companies‡—gave them a joint control over fifteen out of twenty-two important explosives manufacturers other than themselves, and substantial minority interests in the remainder, which in two instances verged on control. In addition, through these twenty-two larger companies or by direct investment, the three had stockholdings in upwards of fifty smaller powder concerns.

Poring over these facts, out of which General Henry du Pont and Solomon Turck had built the Gunpowder Trade Association, Coleman and Pierre grew less and less satisfied with the security it offered to 60 per cent of their own company's assets. As long as Laflin & Rand cooperated, the Du Pont-Hazard investments could be safeguarded by a reasonable degree of control from the home office. However, continued cooperation had become uncertain. Laflin & Rand was dominated by John L. Riker, Henry M. Boies and Schuyler L. Parsons. The first two were nearing the end of their active business life. Parsons had no experience in the explosives business and had never taken an active part in its management. There had been intimations he might sell his interest. It was a question who, a few years hence, would own Laflin & Rand's majority stock.

The uncertainty was most acute in the case of the Eastern Dynamite Company, because it was through this

‡Notably the Oriental Powder Company.

concern that the business, in the opinion of Coleman and Pierre, had greatest promise of future expansion in industrial explosives. They were under no delusions as to black powder. They foresaw that dynamite would push it rapidly into a secondary place. Du Pont-Hazard owned 35.3 per cent of Eastern Dynamite Company stock, Laflin & Rand 29 per cent, and scattered individuals held the remainder. Loss of Laflin & Rand support might, conceivably, permit hostile interests to dictate the policies and management of this important dynamite group, including Repauno, and even force the Du Ponts out of it.

Two courses were open to the cousins: first, to dispose of all the Du Pont-Hazard outside stockholdings, an action that might easily disorganize the industry and jeopardize all they possessed, or, second, to safeguard those holdings by purchasing enough additional stocks of the various companies to give them a majority interest in each.

Coleman du Pont was not one to go backward. "Pierre," he said, "we'll buy Laflin & Rand!"

★ CHAPTER ★
II

Expanding Horizons

UNTIL 1902, New York City was not strikingly different in appearance from any other great city. Then the Flatiron Building rose out of squat surroundings to a height of twenty stories. It symbolized the dawn of a new American era. An era of electricity and whirling motor-driven wheels on which, in time, millions of people were to ride; of pictures that moved and eventually would talk; of machines that could fly; of telephones, radio, X-rays, subways; of typewriters, teletypes, taxicabs and "Trusts."

Between 1898 and 1900, more than forty consolidations took place among the nation's iron and steel companies alone.* Formed were such huge combines as Amalgamated Copper, the larger Standard Oil, American Smelting & Refining, Consolidated Tobacco. In 1901, came United States Steel. Three years later it was stated that six groups controlled 164,500 miles of the country's railroads.

The trend was toward "bigness" in the industrial operations of the country. It led to economies and lower costs that were to establish in the United States during the next quarter of a century the world's highest wage scales and its highest living standards. Large size was to help make

*James Truslow Adams' *History of the United States.*

possible mass production, which, in turn, was to bring many luxuries within reach of the average man.

A product of his time, Coleman du Pont swung boldly into step with it. Bigness made his blood warm, but he was also practical. He felt that so-called bigness of business in itself was not necessarily desirable, that great business size must be justified by serving a constructive purpose in the economic scheme, which smaller enterprise had not and could not fill as satisfactorily. Alfred and Pierre also saw sound reason for buying Laflin & Rand, but making more secure the Du Pont investments was only one of these reasons, and not the most important one either.

Each of the companies in which Du Pont had invested was, in effect, being treated as an independent operating unit. Each maintained its own sales organization, its own branch offices, and purchased its own supplies. Here was a duplication of facilities and overlapping of effort for which, in the end, the consumers of the country paid. Moreover, to make this system of independent operation profitable, it had been found necessary in the past to set up a central governing body—the Gunpowder Trade Association—which contributed little to the common good except coordination in sales practices. The association made no effort to improve its members' products or to create new products through research; and improvement through chemical research was about to become indispensable to the future progress of the industry.

During the century preceding, most of the important innovations in the explosives art—the inventions of nitroglycerin, nitrocellulose, smokeless powder, dynamite, and the blasting cap—had come from European laboratories. Neither the vigorous Coleman nor his thoughtful, quiet-mannered cousin relished being placed permanently in the rôle of followers. They believed that, through some form of organized research, the possibility existed not only of restoring American prestige in the explosives field, but also of expanding the business into fields other than explosives, yet chemically related to them.

It must be kept in mind that Pierre's experience with explosives had been mainly with smokeless powder and guncotton. That work had familiarized him with developments in cellulose chemistry. Too, he had been away from "the Brandywine" long enough to get a broad industrial perspective. Alfred was an expert in black powder manufacture, Coleman was an engineer, but Pierre had been trained as a chemist just at a time when the chemical horizon was expanding and when applied research was becoming more than an academic theory.

Equally, Pierre was by nature an administrator. This, plus his chemistry, made him realize that research such as he envisioned was work that would require organized, consistent effort over a period of years, at a considerable cost in money. Neither Du Pont nor any of the companies in which it was interested could afford to make such a research effort independently, at any rate not on the scale the possibilities warranted; but Du Pont, plus its outside interests and Laflin & Rand, could afford it, if their resources were pooled.

There was still another reason for the purchase of Laflin & Rand. The Sherman Anti-Trust Act was enacted shortly after General Henry du Pont's death. It had cast legal doubt over the complicated structure of the Gunpowder Trade Association and the agreements between its members fixing sales quotas, territories and minimum prices. In 1896, revisions had been made to bring the association into closer accord with the law. But recent court decisions, which had the effect of broadening the scope of the Sherman Act, again raised question as to the association's legal status.

Likewise, it appeared that an agreement between the Gunpowder Trade Association and English and German explosives manufacturers might be considered questionable. This agreement provided, chiefly, that the Europeans would stay out of the United States, Mexico and Central America, and that American explosives companies would stay out of the Eastern Hemisphere.

Thus the cousins found themselves involved in a maze of relationships which were not only uneconomical and difficult to manage, but which were non-constructive and of debatable legality. Yet, if unified, if assembled into a cohesive whole, their scattered holdings could be converted into a highly efficient business instrument capable of raising the plane of progress in the industry to the advantage of everybody concerned, including the public.

In October, 1902, Coleman bought the Laflin & Rand company for $4,000,000, which he paid in bonds of the Delaware Securities Company, a corporation organized for the purpose. Then followed the greatest overhauling the explosives industry in America has undergone throughout its history.

The Gunpowder Trade Association was dissolved. All agreements involving the Du Pont Company and its connections were canceled, the one with the European producers at a cost of $140,000. After a careful survey and inventory of assets, the process of unifying Du Pont interests was begun on a nation-wide scale. Company after company was dissolved and their properties were vested in the Du Pont Company by outright purchase.† Badly situated plants were dismantled. Duplicating sales offices were consolidated at central points. Coleman moved his headquarters from the Brandywine offices into the Equitable Building in Wilmington, where the growing personnel, being assembled from all parts of the country, soon occupied two entire floors.

The character of the powder industry was changed. Gone was the old system of secret controls, under which the real ownership of companies was known only to a few. Ownership and direction were now visible, with a group of vigorous independent companies on one side and the

---

†A new corporation, E. I. du Pont de Nemours Powder Company, was organized and offered to exchange its shares for those of the companies not then wholly owned. The assets of E. I. du Pont de Nemours & Company itself were sold to this new corporation which became the outright owner of all the properties. E. I. du Pont de Nemours & Company remained the holder of a controlling interest of the "Powder Company" until dissolved by order of Court in the year 1912. Financial reorganization again took place in 1915 and the properties became vested in E. I. du Pont de Nemours & Company, the present corporation, and the "Powder Company" was dissolved.

consolidated Du Pont interests on the other in open competition. Coleman established Du Pont price schedules and a sales board at Wilmington that alone had authority to make departures from them. Strict orders were issued against rebates to preferred customers. With the sales rules of the largest producer in the industry thus fixed, a number of new manufacturers entered the field during the next few years.

The independent producers had certain advantages, which they pressed. They could appeal to local pride in home-owned enterprise. Agitation against "bigness" was in their favor, especially when the customer himself owned a small business. Moreover, they soon learned they could undercut the Du Pont prices with impunity. Operating on a national scale, Coleman could not afford to involve himself in local price wars without jeopardizing his entire price structure.

Bigness also had advantages, however. The Du Pont research facilities were augmented. At Repauno, the Eastern Laboratory was established in 1902 under the direction of Dr. Charles L. Reese, a noted industrial chemist. Dr. Reese was a graduate of the University of Virginia and of Heidelberg, had taught chemistry at Johns Hopkins, and had served as chief chemist for the New Jersey Zinc Company. Surrounding himself with able assistants, he launched a research organization specializing in explosives that was destined to make notable contributions to the art in its industrial applications. Outstanding were the developments of low-freezing dynamites and of the so-called "permissible" class of explosives, or those that can be safely used in gaseous and dusty mines.

In 1903, a second research laboratory called the Experimental Station was authorized and soon established in its own building on "the Brandywine." The move was significant of the broadened vision of the management. The station was placed under the ægis of a special new division of the company, the Development Department, which in turn was put in charge of Coleman's old mentor, Arthur J. Moxham.

Coleman had long admired Moxham's organizing genius, his foresight and business acumen. Now he entrusted him with the task of pointing the Du Pont Company into fields beyond explosives. The Development Department and its research auxiliary were to investigate new outlets for the products of Du Pont factories, existing or potential, and also establish independent sources of Du Pont principal raw materials that would relieve the company of its then total dependence upon others in this respect.

Moxham's instructions to his researchers were explicit. Henceforth they were to regard Du Pont not as a producer of explosives alone, but as a chemical manufacturer ready to venture wherever its logical chemical interests might lead. The diverse fields related to explosives were to be explored, the ways charted for a vigorous expansion therein. Four such fields were obvious. Solutions of nitrocellulose, the base of Du Pont smokeless powder, were being used in the manufacture of lacquers for coating metal hardware. In still other forms, nitrocellulose was being employed in the making of "artificial leather," the pyroxylin type of plastics, and photographic film.

Shortly, upon the Development Department's recommendation, the International Smokeless Powder & Chemical Company was purchased. At its plant at Parlin, New Jersey, this company produced not only gunpowder but nitrocellulose solutions and solvents as well. It had a thriving little "brass lacquer" business, some excellent chemists and other trained personnel. This step, taken in 1905, was the first in a program of development "beyond explosives" that eventually was to embrace activities of tremendous complexity, yet all chemically interrelated.

A Technical Section was organized in 1905 as an adjunct to the Explosives Sales Department. The purpose was to have highly trained technical men supplement the explosives salesmen in rendering service to customers with special problems, and to teach the safest, most effective and economical methods of putting explosives to work. Through efforts of this kind, large sums were to be saved quarrymen, mine operators and construction engineers. Blasting in its intricate phases was to be raised to a science.

By 1906, the company was expending $300,000 annually on research. Such an outlay, then exceeded by few other industrial concerns, was justified by the consolidated and enlarged operations, but under other circumstances it would have been impossible for any period of time.

The greatest gain, of all, however, was in man-power. The ablest men in the explosives industry not already in the Du Pont ranks were drawn into Coleman's organization by the consolidation of companies. One was Amory Haskell, who had made a brilliant record as president of Laflin & Rand and who was probably the foremost executive in the high explosives field. Another, Colonel Edmund G. Buckner, was equally talented as a salesman. Kentucky-born, and with an early experience in banking, Buckner had been president of the Marsden Company, organized to develop chemical uses for cornstalks, and later of the International Smokeless Powder & Chemical Company. Where Haskell stood in the world of industrial high explosives, Buckner was destined to stand in the world of military smokeless powders.

From the International company also came two brilliant chemists, Harry Fletcher Brown and Charles F. Burnside. A Harvard graduate, Brown had spent seven years at the United States Navy's Torpedo Station at Newport, Rhode Island, during which time he had been intimately identified with the Navy's exhaustive experiments with smokeless powder. At first chief chemist, and then superintendent of the International company's smokeless powder operations, there was no man in America better informed than he on military powder development. Burnside had entered the powder business by way of Ohio University, the Corcoran Scientific School and the Dittmar Powder Company, founded by Carl Dittmar, an early associate of Alfred Nobel. He had studied under Dr. Charles E. Munroe, a famous American explosives authority, and for several years had been closely associated with Brown.

Another invaluable addition was Major William G. Ramsay, who was chief engineer of the Eastern Dynamite Company when that business passed to Du Pont. A Californian, he had learned engineering at the University of Virginia and in practical railroad work, followed by military service during the Spanish-American War. Major Ramsay was placed at the head of the Engineering Division of the consolidated Du Pont companies. This division was to design and build all Du Pont plants and, later, was to be of stupendous international importance during the First World War.

Frank L. Connable became assistant to Alfred I. du Pont in the management of plant operations. He had made a notable record as general manager and then president of the Chattanooga Powder Company, but he was to exceed that record in the Du Pont service. Amory Haskell brought with him a trusted associate, Charles Copeland, a native of New Jersey and Harvard-trained, whose previous experience as treasurer of the Eastern Dynamite Company and of the Lake Superior Powder Company was to be of help in administering the intricate financial details of the new expansion program.

The old Du Pont Company and its affiliates, as well as the ranks of every company absorbed by purchase, were combed for men of vision and talent. Charles L. Patterson, pioneer Repauno dynamite salesman, had been one of the first men in the industry to advocate that industrial explosives should be made specifically for specific needs, and sold by the manufacturer's own sales experts instead of by agents. He was given the opportunity to put this Repauno sales plan into effect on a national scale, which entailed a reorganization of most of the company's sales facilities and an augmented research program in the Du Pont laboratories. The fact that his chief assistant, later his successor, in sales, was William Coyne, whose basic training had been with problems that involved the shipment of explosives, was evidence of an important new policy in Du Pont management. The aim now was to get cooperatively formed views and decisions. Organized effort was the new keynote of all action.

An Executive Committee was formed. Arthur J. Moxham was appointed to it. Hamilton Barksdale was a member—his prominence in the new management was even greater than in the old. Similarly, Amory Haskell was included. Other places were taken by Alfred I., Pierre and Francis I. du Pont, who was the eldest son of Francis G. As president, Coleman served ex officio.

Inauguration of this committee, in 1903, ended the century-old precedent that Du Ponts alone should exercise official authority in the present company. Another precedent was broken in admitting to it Francis I., who was not yet thirty, although a chemist who was showing unusual inventive talents. And still a third precedent was shattered when bonuses in the form of Du Pont stock were awarded Barksdale, Amory Haskell, Moxham, and later others, which made them share owners of the business. Members of the Executive Committee were vice-presidents and also members of the Board of Directors.

Henceforth this committee was to guide every major decision of the company. Its power was to grow. It was to become the administrative agency of the Du Pont business.

With eyes to the future, Coleman drew into the company more young Du Ponts or encouraged the family members already there. A. Felix du Pont, second son of Francis G., was made an assistant superintendent at the Carney's Point smokeless powder plant. Eugene du Pont, Jr., became Assistant Director of Sales of the Central Division. Among Pierre's younger brothers, Irénée was summoned from the Manufacturers' Contracting Company, Lammot from the Pencoyd Iron Works, and William K. was assigned to join his cousins at Carney's Point. Other recruits were "Dr." Alexis du Pont's son, Eugene E., and the late Eugene's other son, Alexis Irénée, third of that name in three generations.‡ Coleman believed that the more Du Ponts he could put into the company hopper, the more likely was it to shake some Du Pont of real ability to the top.

The company's aggressiveness, the reports that it was "going places and doing things" also drew to its ranks many young men of promise in no way identified with the Du Pont family. They came, most of them, fresh from the technical schools. The success of Coleman's policy in developing leaders is apparent when one runs through the roster of present-day high Du Pont executives. Many

‡Of the younger Du Ponts mentioned here, A. Felix, Irénée, Lammot, Eugene, and Eugene E. were all serving as members of the company's Board of Directors on January 1, 1942. William Kemble du Pont, born in 1875, died in 1907; Alexis Irénée du Pont, 3rd, born 1869, died in 1921.

of them came into the company between 1902 and 1909.

Two major contributions to the company's human relations were voted by the new management. A fund was established in 1904 for pensioning employees of fifteen years or more service, who, because of age or infirmities, became incapacitated for further work. The following year a bonus plan was formalized, following the line of thought Amory Haskell had been pursuing in the Repauno company for eight or ten years. The formalized plan, still in effect, provided for two classes of bonus to employees in the form of the company's stock: Class "A" bonus for conspicuous service of any nature, as, for example, inventions of fundamental importance, or suggestions and work which importantly affect the company's welfare and progress; and Class "B" bonus to those who contributed most in a general way to the company's success by their ability, efficiency and loyalty. With the company's rapidly increasing earnings, the bonus system was to make it possible for men of unusual ability to build estates comparable to what they might have built as executives in their own businesses.

The company was in a broadening field of industrial action. Its leadership was vigorous. Behind it was a century of experience and service that had won distinction and respect. An *esprit de corps* grew through its rank and file that pointed to great accomplishments. The newly built organization strained at its leash. There can be no greater achievement by management than to create this spirit in a company.

United States vs. Du Pont

AMID the harmony at Wilmington, a discordant note sounded, however. For more than twenty years Robert S. Waddell had been a sales agent for the Hazard Powder Company. Stationed at Cincinnati, he had experienced some of the keenest competition of any one in the powder industry. Recognizing Waddell's resourcefulness and long experience, Coleman du Pont had called him to Wilmington and appointed him general sales director.

Waddell was an individualist who had been restless under the supervision of the Gunpowder Trade Association. Coleman's stricter sales policies, his consolidation of sales offices and his rule against rebates and price-cutting, heightened Waddell's disgust at the trend the industry was taking. He announced that he was going to start a black powder mill of his own.

Coleman offered to help him. A tentative contract was drawn up in which he agreed to provide 49 per cent of the capital if Waddell supplied 51 per cent. The latter was to name three directors of the new company, Coleman two, and Waddell was to be president at a salary of $10,000 a year. But when the two failed to agree on a plant site, Waddell repudiated the contract and resigned from the Du Pont Company.

Forming the Buckeye Powder Company, in 1903, he erected an independent mill at Edwardsville, Illinois, despite Du Pont efforts, as he charged later, to stir up the community against him and to prevent him from getting machinery. Two years of profitable operation followed, but then all business declined incident to the depression of 1907. Without reserves sufficient to carry him, Waddell offered his property for sale. He said the mill had cost him $118,000, but he sold it for $75,000 to F. W. Olin and Almon Lent, presidents respectively of the Equitable Powder Manufacturing Company and Austin Powder Company. Olin spent another $100,000 on the mill, changed the name to Western Powder Company, and brought it back into successful operation.

In the meantime, Waddell had been savagely attacking the "Powder Trust" in letters to congressmen and newspaper editors. While in the employ both of Hazard and Du Pont, he had made copies of all letters dealing with the activities of the Gunpowder Trade Association in its efforts to maintain prices and control competition. He now took these documents to Washington and presented them to the Department of Justice. The department acted on July 7, 1907, by instituting suit against the Du Pont Company and its officers, its undissolved subsidiaries and the former members of the Gunpowder Trade Association for violation of the Sherman Anti-Trust Act. Waddell followed with a suit of his own in which he charged that his failure in the powder business was the result of a Du Pont conspiracy, by reason of which he had been injured to the extent of $404,428.07. He asked for three times that sum in damages.

Waddell's private suit, tried in the Federal District Court at Trenton, New Jersey, was not finally closed until 1914. The jury then decided Waddell had no cause for action and the defendants were exonerated. In the suit of United States *vs.* Du Pont, however, though it was based mainly on the ex-sales manager's charges, another Federal court—in Delaware—decided in favor of the prosecution.

The Government charged that the Gunpowder Trade Association and later the enlarged Du Pont Company both had been organized to stifle competition in the explosives industry. It alleged that in 1907 the Du Pont Company and its subsidiaries controlled, of the total explosives business in the United States:

Black blasting powder.. 64%
Soda blasting powder.. 72%
Black sporting powder... 74%
Dynamite 72%
Smokeless sporting powder.................................... 64%
Smokeless military powder....................................100%

The Government contended this constituted a dominance of the industry that was, in effect, a restraint of interstate trade.

Hearings were held in Wilmington, New York City and Chicago. Hundreds of witnesses passed to and from the stand. Their testimony filled thirteen volumes. The case dragged over months, then years. And still the lawyers asked questions. Still the parade of witnesses continued.

Almost every person of importance in the explosives industry, not to mention minor figures, testified for one side or the other. The ashes of the dead Gunpowder Trade Association were raked and reraked, musty ledgers were reopened, almost forgotten tales retold as the old battles of the industry were fought once again in volumes of words. James Scarlett, chief of Government counsel, emphasized that Coleman had not dissolved the Trade Association until after Du Pont had bought Laflin & Rand. The case hinged on that contention.

On June 12th, 1911, after four years of litigation, the court declared the defendants in violation of the Sherman Act. The following is from the opinion:

It is a significant fact that the Trade Association . . . was not dissolved until June 30th, 1904. It had been utilized until that date by Thomas Coleman du Pont, Pierre S. du Pont and Alfred I. du Pont in suppressing competition and thereby building up a monopoly. Between February, 1902, and June, 1904, the combination had been so completely transmuted into a corporate form that the Trade Association was no longer necessary.

Consequently, the Trade Association was dissolved and the process of dissolving the corporations whose capital stocks had been acquired, and concentrating the assets in one great corporation, was begun. . . . The proofs satisfy us that the present form of combination is no less obnoxious to the law than was the combination under the Trade Association agreement.

Sixty-four corporations had been acquired and dissolved by Coleman. Sixty-nine others were in process of being dissolved when the suit interrupted. The thousands of stockholders of these companies were well satisfied with the deals that had been made—their Du Pont stock was paying excellent dividends. The situation was unprecedented. Frankly admitting the impossibility of restoring original conditions in the industry, the court issued an interlocutory decree directing that the Government and company jointly should work out a plan for the company's reorganization, subject to the court's approval.

At this point came a third unprecedented happening. Only in military smokeless powder had the Government attempted to establish the fact of a 100 per cent Du Pont monopoly. Such a monopoly had been freely admitted by the company in so far as private manufacture was concerned. It seemed certain, therefore, that the court would insist above all other considerations that manufacture of military smokeless powder should be opened to competition. Both the Army and the Navy, however, protested against such action.

The chiefs of the ordnance bureaus, the president of the Joint Army and Navy Board on Smokeless Powder, and other high officers appeared before the court. In peace and war, they testified, the nation's transactions with the Du Pont Company had been satisfactory over more than a century. The fact that the Government operated powder plants of its own, under competent officers, they argued, provided it with effective means of gauging the fairness of the Du Pont smokeless powder prices. Moreover, they added, the Du Pont plants and laboratories had always been open to Army and Navy experts, with the result that both services had benefited by Du Pont patents and secret processes for use in the Government's own powder plants. Instances of cooperation were cited that had saved the United States large sums of money—as a rule, without compensation to the company.

Agitation in Congress had continued throughout the trial with little or no regard to the evidence. Ignoring it, the court handed down a final decree, June 13th, 1912, embodying the statement that a division of the military powder business among several competing companies "would tend to destroy the practical and scientific coop-eration now pursued between the Government and the defendant company, and to impair the certainty and efficiency of the results thus obtained." The decree added, "No benefit would accrue to the public by dividing this business between several competing concerns, while injury to public interests of a grave character might and prob-ably would result therefrom."

In the light of future events—those of the World War, then unforeseen either by court, Government or company —this decision was to outweigh in importance the court's main decree ordering a dissolution of the Du Pont indus-trial interests, which at that moment was seemingly the major issue involved.

As of December 15th, 1912, following a plan suggested by the Government and approved by the court, the Du Pont business in industrial explosives and sporting powders was divided among three companies—Du Pont, and two new independent corporations, the Hercules Powder Company and the Atlas Powder Company. The new companies were named after the two famous brand names of dynamite that were assigned to them. Laflin & Rand, the Eastern Dynamite Company, the International Smokeless Powder & Chemical Company and the numerous smaller Du Pont subsidiaries were ordered dissolved forthwith.

The larger of the two new companies, Hercules, was given eight black powder mills, three dynamite plants and the Laflin & Rand patents for the manufacture of smokeless sporting powder. Its authorized capital was $6,500,000 in common stock and a like amount in 6 per cent income bonds, with which it paid for the assets it acquired. The Atlas Powder Company was assigned six black powder mills and four dynamite plants, for which it paid $3,000,000 in common stock out of an authorized $5,000,000 and additionally $3,000,000 in bonds.

The Du Pont Company was left with twelve black powder mills, five dynamite plants and three plants for the manufacture of military and sporting smokeless powder under its own patents. As these assignments included the substantial developments of Repauno, Carney's Point, Mooar, and the Brandywine mills, Du Pont continued to be the largest producer in the industry and, as noted, the only private producer of military smokeless powder in the United States.

The Hercules and Atlas companies, however, were substantial corporations. Experienced managements were placed over them and they quickly established their independence. The Hercules company early exchanged its outstanding bonds for preferred stock to strengthen its financial position. Atlas acquired control of the Giant Powder Company, Consolidated, on the Pacific Coast. Authorized by the court's decree to utilize for a period of five years the Du Pont Company's engineering, chemical and purchasing departments, the new companies rapidly improved and added to their properties. Later, they established their own chemical research laboratories. Today, as diversified chemical manufacturers as well as producers of explosives, Hercules and Atlas rank among Du Pont's most aggressive and respected competitors.

Thus ended the "Powder Trust."

★ CHAPTER ★
IV

Du Pont vs. Du Pont

IN JANUARY, 1913, after the division ordered by the court, the assets of the Du Pont Company were worth more than six times what Coleman, Pierre and Alfred had paid for the company in 1902. The few clerks who had served Eugene du Pont had increased to a headquarters staff of a thousand. Four times that number of plant workers were employed. The Du Pont Building, which housed the home offices, loomed over Wilmington's business center. Its first section comprising 105,000 square feet of floor space was completed in 1907. Additions more than doubling that space were made by 1913. A 300-room hotel and a theater were housed within the building's four walls.

Du Pont industrial explosives in 1913 were blasting the way for some of America's greatest engineering accomplishments. One, the Panama Canal, under construction since 1904, was to consume 61,000,000 pounds of dynamite. All the nation's wars up to that time had not required the explosive energy devoted to this one huge task of linking the Atlantic and the Pacific.

Yet the canal was dwarfed by the extraordinary expansion in mines, quarries and construction going on at home. The New York City subway system, a wonder of the modern world, had been in the building uninterruptedly since 1900. Grand Central Terminal, with its labyrinth of underground stations and tracks above which rumbled the city's traffic, and the Woolworth Building, were nearing completion.

Millions were being spent to shorten railway lines and reduce grades by means of tunnels, bridges and cut-offs. In progress were vast land reclamation projects, hydro-electric dams, such water-supply systems as the Catskill and the Los Angeles aqueducts. Almost 300,000,000 pounds of dynamite and 250,000,000 pounds of black powder were being used annually in the United States for industrial purposes.

The old "rule-of-thumb" was gone. Chemical science now governed the explosives industry. The new order was reflected in the character of the Du Pont personnel and especially in the effort being made in the Du Pont research laboratories. To facilitate their administration in 1911, a Chemical Department had been inaugurated. Doctor Reese was appointed Chemical Director for the company as a whole. In line with the broader perspective, the Development Department, too, had prompted a second major step toward a more diversified chemical manufacture. In 1910, the Fabrikoid Company of Newburgh, New York, had been bought outright and Du Pont prefixed to its name.

"Fabrikoid" was the trademark used for a new waterproof fabric made by treating cotton cloth with successive coatings of a pliable nitrocellulose lacquer. It had been sold since 1895 as an artificial leather. Du Pont chemists had found ways to improve this product. Acquisition of the Fabrikoid Company's plant and patents had provided immediate manufacturing facilities, as well as the nucleus of trained men around whom could be built an enlarged organization. This method of expanding into new lines through the purchase of "going concerns" was to become typical of much of Du Pont's future development.

The improved Du Pont lacquered fabric was tough, pliable and resistant to grease, oil, perspiration, mildew, and other agents that cause leather to deteriorate. It could be produced in a variety of weights, colors and surface finishes at a fraction of leather's cost. In the first flush of an expansion that soon was to put millions of lower-priced open cars on the roads, the automobile industry needed such a material for upholstering seats and covering decks, luggage, and other parts of cars subject to hard wear, dirt and exposure. No material wholly suitable for these purposes had been available previously. Other chemically conceived fabrics of the class shortly were to enter the field in force and, through steady further improvement, demolish one of the first formidable barriers to mass production of motor cars. Thus Du Pont became a leader in a development wherein the possibilities for future accomplishments were limited only by human ingenuity, and the competitive incentive to betterment was keen and ever-growing.

Looking upon his work, Coleman du Pont was satisfied. Under his leadership the company had come to a crossroads and boldly taken the way toward the rising sun. It was a well-integrated organization, strong in the union of many able men. Within the company's own incubators of research even now were being hatched the means of continuing progress. The Development Department was scouting broader fields.

To Coleman, who fretted at routine and the details of daily administration, his greatest contribution to the company had been made. New adventures beckoned him. He had already participated in a syndicate that had built the McAlpin Hotel in New York. Now he plunged, alone, into the biggest office building project of those pre-war years, the forty-story Equitable Building. The site, at 120 Broadway, between Pine and Cedar Streets, was one of the most valuable in Manhattan. Thereon Coleman proceeded to build a structure that was to have 2,300 offices sufficiently spacious to accommodate 15,000 occupants. The cost, with the site, was to be $30,000,000. At the same time, with Lucius Boomer, manager of the McAlpin, he discussed forming a chain of hotels along the Atlantic Coast with the McAlpin and Waldorf-Astoria in New York, the Bellevue-Stratford in Philadelphia, and the Willard in Washington as its principal links.

Wilmington saw less and less of him in those bustling days of his broadening outside interests. Then, suddenly, matters were complicated by his growing ill health. For two years, unknown except to a few intimates, he had been under the care of a physician. In 1906, the Executive Committee had asked Pierre du Pont to act as chairman during Coleman's absence, and in January, 1909, at Coleman's request, the Committee definitely delegated to Pierre the duties of president. Later in the year Coleman offered his resignation, but, not wishing to lose him, the Executive Committee persuaded him to accept, instead, an indefinite leave of absence.

The reason that Pierre, not the older Alfred, was selected to take up the president's burden was that the loss of the sight of one eye in a hunting accident and an increasing deafness had led Alfred to resign from the Executive Committee two years earlier. Frank L. Connable, Alfred's chief assistant, succeeded to the vacated membership. Thus, in 1909, Pierre alone of the three cousins was active in the daily administration of the company's principal affairs.

The situation had important consequences. The uncertainties concerning the presidency greatly strengthened the already powerful position of the departmental heads and of the Executive Committee. The latter became the company's administrative manager-in-chief. The president, or acting president, served only as the chief executive officer. He was expected to lead in recommending changes in policy or in practical procedure, but decisions became permanently an Executive Committee prerogative. The policy of cooperative action launched by Coleman thus became even more decidedly the keystone of Du Pont management.

Coleman's prolonged absence, Pierre's absorption in the organization, the effect of Alfred's growing deafness, all

tended to separate the interests of the three cousins. As the
year of 1914 opened, Pierre's most intimate company
associates had become inevitably the members of the
Executive Committee, heads of departments and. other
executives with whom he was in regular contact as acting
president. Alfred emerged now and then to discuss financial and other problems of the business, but his concern
was centered in the black powder manufacturing department. Coleman's health seemed better. He was, indeed,
feeling well enough to resume his place at the company's
head. He shortly did.

Then came June of 1914. The Austrian Archduke Francis Ferdinand and his Duchess were assassinated by a
Serbian fanatic at Serajevo. One month later Europe was
at war! And presently, as in all other wars of that continent over a century, a stream of wealth began to flow
westward over a now submarine-infested ocean in return
for American wheat, beef, cotton, woolen goods, lumber,
coal, gunpowder and other supplies.

The stream grew swiftly in volume. A rivulet from it
began emptying into Wilmington. It brought effects of
which nobody there had dreamed. One of these, within
another year, was to change materially the once close
business relationship of Coleman, Pierre and Alfred
du Pont.

On one or two occasions Coleman and Pierre had discussed the desirability, as Coleman put it, of rewarding
more generously the principal executives of the Du Pont
Company. He felt the regular bonus plan adopted in

1905 was not enough, that those now most responsible for the company's management should, as in the cases of Amory Haskell, Arthur J. Moxham, Hamilton Barksdale and others earlier, share more largely in the company's ownership. He had in mind chiefly the younger men who had come up over the decade to the Executive Committee and major managerial posts, and who had not shared in the special stock distributions made after Coleman had assumed the presidency.

Both Coleman and Pierre knew that able men could not be held long in any company, if the opportunity offered them was not as great as their prospects elsewhere. The success of the whole Du Pont managerial program was grounded on developing and retaining strong executives who would regard their work with the company as a career, the roots of which were anchored deep in the company's soil.

Coleman's health again became bad in the late summer of 1914. He underwent an emergency operation. December found him facing the possibility of a yet more serious one. He formally proposed that the company purchase a large block of his Du Pont Common Stock, at the market price, for distribution among the company's "leading men."

He wrote to Pierre:

This ownership of common stock by the leading men is of so much importance to the company that, as a member of the Finance Committee (the company having cash beyond its requirements), I would recommend the funds needed to carry the stock for these men be advanced by the treasury.

Pierre reported Coleman's plan to Alfred. And a week later, before leaving for the Mayo Clinic at Rochester, Minnesota, Coleman wrote further:

I have given a great deal of thought to my letter of December 7th as to distribution of this stock, and my judgment is that each member of the Executive Committee be allowed 1,500 shares. That the men whose salary is $500 a month and over will be allowed to subscribe for three times the amount of their yearly salaries. This makes a total, according to the memorandum you left with me, of 20,700 shares.

That same day Pierre was handed another letter. It was from Alfred and read:

After giving the question of purchasing a large block of Du Pont Common further consideration, I believe that 160, the price you talked of, is too high. 150 would be purchasing it on a 6 per cent basis, and 160 a 5 per cent. This is too low a rate for the company to invest its spare funds, and, furthermore, if it were offered to any of our employees, it would not be sufficiently attractive at that price. I see no objection to the company purchasing this stock, but the question of price is one of grave importance.

A few days later Alfred corrected his figure of 150 to 133, and added that he questioned if investment of the company's surplus earnings "on a basis even as high as 6 per cent could be justified." Still later, he suggested the price of $125 a share.

The company's Finance Committee met December 23rd. Present were Alfred, Pierre and William du Pont—the same William of early Repauno days. While always interested in the company, he did not take an active position, following his resignation as Repauno's president, until he became a member of this important committee. The only absent member was Coleman. The official record of what transpired was set down by the meeting's secretary, L. R. Beardslee, as follows:

Mr. P. S. du Pont presented a letter from Mr. T. C. du Pont offering to sell 20,700 shares of the common stock of this company at $160 per share. After discussion, it was moved and carried (Mr. P. S. du Pont voting in the negative) that Mr. P. S. du Pont be instructed to advise Mr. L. L. Dunham, attorney* for Mr. T. C. du Pont, that we do not feel justified in paying more than $125 per share for this stock.

Pierre reported this action to Lewis L. Dunham, but did not get in contact with Coleman until after the New Year had passed and some strength had returned to the sick man, whose second operation had taken place. Then, on January 4th, 1915, Pierre counseled that the matter be left in abeyance for the moment. "I am sorry and provoked," he wrote, "that the proposition did not go through, for I feel your offer was a generous one and should have had more considerate treatment; but, like many other things, the final result cannot be obtained quickly."†

Pierre added that since the start of the war in Europe the Du Pont Company had received orders from the Allied Powers for 35,000,000 pounds of smokeless powder. It was one of many informative letters that went forward to Coleman from Pierre during ensuing weeks.

*Confidential secretary to Coleman du Pont, with power of attorney.

†A proposal similar to the one made to the Du Pont Company was also submitted by Coleman to the Hercules Powder Company and to the Atlas Powder Company, in both of which he held stock that had come to him in the dissolution proceedings resulting from the antitrust suit of 1907–11. Both of these companies accepted the proposal.

The war had come to Wilmington. Fighting against Germany's superior military preparedness, France, Great Britain and Russia were making frantic efforts to augment their own deficient powder supplies. Du Pont was the one important manufacturer of smokeless powder to whom they could turn and expect large production within a reasonable time. Something of the situation the company faced was reported by Pierre to Coleman in a letter of January 9th, 1915:

The march of the smokeless powder business is going on uninterruptedly. In order to take care of deliveries for the latter part of this year, we have now increased our manufacturing capacity to 6,000,000 pounds of guncotton per month. The powder manufacturing capacity is not quite equal to this, but a new negotiation will probably make it necessary to round out the 72,000,000 annual guncotton capacity to a similar amount of powder. . . .

The men of the Engineering Department have responded splendidly to the heavy calls made upon them. The first extensions have come in ahead of time and we are now working at what would have seemed an enormous output a few months ago; though it is not nearly half of what we expect to do. . . .

The letter went on to say that because of the absence of William du Pont, the subject of the stock purchase had not been further discussed, but that it was Pierre's judgment a favorable decision would be reached upon William's return and reconsideration of the offer then.

From his bed in St. Mary's Hospital at Rochester, Coleman had already written to Pierre in his own hand, still shaky, on the Finance Committee's action. He questioned Alfred's motives, instructed Pierre to withdraw the stock sale offer if that action seemed to him to be in the company's best interest. Later, he himself withdrew his proposal. On January 16th, Du Pont Common was quoted at 182 on the New York Stock Exchange.

The pending disposal of Coleman's stock had been rumored for some weeks in Wall Street. The matter now became of international concern. Sometime during the fall of 1914, J. Amory Haskell had been approached by Captain Fritz von Rintelin, on behalf of German interests, to determine if the Du Pont Company could not be bought outright. The idea had been immediately and summarily rejected by Pierre du Pont, but it was probably responsible for reports reaching London and Paris that, with Coleman's stock purchasable, control of the Du Pont Company was in danger of falling into German hands. Still another report whispered in London was that the company was overextended, could not possibly fill its orders, and faced bankruptcy. Alarmed, the Allies dispatched a confidential agent to New York. He was met by Pierre, his brother Irénée and Colonel Buckner, who was the company's representative in military powder transactions. Pierre reassured the agent. Nothing to cause alarm had happened. Pierre pledged his word it would not happen, and forthwith reported the incident to Coleman, Alfred and William in a communication to the Finance Committee.

Belatedly, on February 16th, Alfred wrote Coleman an explanation of his own and William du Pont's position, repeating the stock valuation of $125. Coleman's nettled answer was the question: "Won't you please advise me how much of your common stock you are willing to let go at this time to the important employees at the price suggested by you, $125 per share? Probably I can join with you in an offer."

Alfred did not reply.

Du Pont Common had reached 206 a few days earlier. On the company's books were orders for more than 61,000,000 pounds of explosives from the Allied Powers. At Hopewell, Virginia, a great new guncotton powder plant had started operation. There was talk of a 12 per cent dividend for 1915.

Prophesying the stock would go to 300 before the year ended, Coleman, still in Rochester, authorized his secretary, L. L. Dunham, to negotiate for the sale of 20,000, 30,000 or even 40,000 shares of his Du Pont Common Stock at $200 for the benefit of "the men who are now at the helm and actually doing things."

"If much time is desired," Coleman warned Pierre, "the price should be higher."

Dunham immediately conferred with Pierre and on February 17th telegraphed Coleman as follows:

"Wire me if you would be willing to sell 40,000 shares and pool balance for voting purposes. Another proposition: would you be willing to sell entire holdings, both cash."

The answer was that Coleman would sell *all* of his stock at $200 a share for the common, but that he would not pool any part if retained by him unless he knew the conditions of the pooling agreement.

Hurriedly, Pierre called together a group of close associates. Coleman's holdings totaled 63,214 shares of common stock and 13,989 shares of preferred.

Three days later, a syndicate composed of Pierre, Irénée, Lammot and A. Felix du Pont, John J. Raskob and R. R. M. Carpenter agreed to buy all of Coleman's common stock at $200 a share and all of his preferred stock at $85 a share. Payment was to be $8,000,000 in cash, $5,831,865 in 6 per cent seven-year notes. Coleman wired his acceptance.

Reports of the transaction were carried in the newspapers of February 28th upon its consummation. Alfred insisted the stock should have been bought for the company; William wired from Georgia he presumed the purchase had been made for the company, that he would consider any other action breach of faith.

However, the syndicate had not acted for the company. Theirs had been the only authority. Coleman's offer to the company had been rejected and withdrawn. The syndicate owned the stock.

Pierre contended the company's credit had not been used, that the purchase had been a personal transaction, made after the company had failed to accept Coleman's offer. He said a loan of $8,500,000 had been negotiated through J. P. Morgan & Company, secured by collateral put up by the syndicate, not the company.

"Then you refuse to turn over the stock?" asked Alfred.

Pierre did refuse. The stock belonged to the syndicate. In line with Coleman's wishes, the principal men of the company would be given the opportunity to take parts of it—in fact, they had already been approached.

Next day, however, Pierre decided to consider an offer from the company to buy the stock, if it wanted it, subject to a reservation of some 8,200 shares which he had already set aside for company executives. Eighteen directors met in special session, on March 6th, to consider the matter and to elect a successor to Coleman, who had submitted his resignation as president.

Pierre told the directors he could not continue with the company if he lacked the confidence of the board. The answer he got was his unanimous election to the presidency. Action on the stock proposal was taken at a second meeting on March 10th, when a resolution to buy the stock at $200 a share was defeated, only Alfred, William and Francis I. du Pont voting affirmatively. The governing reason was the opinion of J. P. Laffey, the company's chief counsel, that legally the stock could be purchased only out of surplus funds, and the surplus was about $5,000,000 short of the sum required.

Here, it is pertinent to observe the men who made up the syndicate.

Pierre's brothers, Irénée and Lammot, joined the company in 1902. They had risen rapidly. Irénée, the elder, was elected in 1904 to the Board of Directors; in 1908, he was made manager of the Development Department; on February 15th, 1914, he had become a vice-president. Lammot had served in turn as division manager, division superintendent and general superintendent of the Black

Powder Department. A. Felix du Pont had spent nine years at the Carney's Point smokeless powder works, four as superintendent. Since 1911, he had been technical assistant on small-arms powder to Harry Fletcher Brown.

John J. Raskob was born in Lockport, New York, in 1879. The Lockport public schools and business college furnished his education. A local pump manufacturing concern gave him his first work as a stenographer. Pierre engaged him in 1900 as a secretary and quickly discovered the young man was much bigger than the position he held. In a year, Raskob was treasurer and financial manager of the Dallas, Texas, street railways. Joining the Du Pont Company when Pierre did, his sense of financial management speedily matured. In 1911, he became assistant treasurer of the company. Two and a half years later, at the age of thirty-five, he was treasurer.

R. R. M. Carpenter was born in Wilkes-Barre, Pennsylvania. He studied architecture at the Massachusetts Institute of Technology, where, as a fellow student, he met Lammot du Pont. Six years of varied business experience in eastern cities found him, in 1904, serving the Manufacturers' Contracting Company, a Du Pont subsidiary, as treasurer. In 1906, he joined the Purchasing Department of the parent company and two years later transferred to the Development Department, becoming its manager in 1911.

The financing of the purchase of Coleman's stock was made through the Christiana Securities Company, a corporation organized for the purpose, to which the six syndicate members themselves contributed a large block of stock of E. I. du Pont de Nemours. These six moved immediately to distribute a part of the Securities company stock to the Du Pont Company's principal department heads—Harry Fletcher Brown of Smokeless Powder, Harry G. Haskell of High Explosives, Major William G. Ramsay of Engineering, William Coyne of General Sales, Frank G. Tallman of Purchasing, and John P. Laffey, head of the Legal Department. However, whereas all earlier discussions of the matter had contemplated giving the "leading men" merely the opportunity to purchase stock with the company's financial assistance, the distribution as actually made by the syndicate was in the form of outright gifts, subject only to the condition that each recipient should remain in the company's employ for the next year, and to the Securities company's indebtedness.

This was a far more generous action than the company itself could have taken had it become the purchaser of the stock, and at one stroke assured two things: retention during the war emergency of men highly important to the success of the company's unprecedented program, and their self-interest in that success as part-owners of the company. In other words, the principle of partnership, which involves the sharing of results as well as responsibilities, that the old firm had exemplified and the bonus system of 1905 had reaffirmed, was now re-emphasized in striking fashion at a time when any considerable disaffection in the company's leadership might easily have caused disaster. Moreover, the precedent of partnership was buttressed for the future, with the result that to this day the Du Pont Company is in the happy condition, rare in large concerns, of having the majority of its stock directly represented on its Board of Directors, in addition to having as members of this board the working heads of the organization's executive management. The expansion and elaboration of the original bonus plans have paralleled the company's growth almost as a matter of course, extending the partnership feeling throughout the organization.

The department heads who benefited individually from the syndicate's gifts of stock were all in "key" positions of responsibility in that pivotal winter of 1915, when some worried observers felt that Du Pont might have undertaken a task much too big for it. Frank G. Tallman, a native of Iowa, had been in charge of the company's purchasing offices since 1905, and now at fifty-four was widely known and highly competent in his specialized field. John P. Laffey, another Midwesterner, had joined the Du Pont legal staff in 1903 to become its chief in 1913, after ten years' experience in corporate legalities more complex than most lawyers encounter in a lifetime. The younger Haskell, of whom we last heard as secretary of the Repauno Chemical Company, had made a distinguished record in the operating end of high explosives manufacture. His organization, in 1896, of that department of the business had led to his becoming its general manager, and the rapidly growing demands of the World War had already made him the No. 1 manager of high explosives operations in the United States. Harry Fletcher

Brown and Major Ramsay had risen to positions in their respective fields—smokeless powder and explosives plant construction—that were correspondingly important. Indeed, the success of the Du Pont manufacturing expansion underway in all three of these fields, in 1915, spurred by the emergency demands of the Allies, depended largely on the leadership of this trio of executives, and upon Tallman's ability to supply them with materials.

Meanwhile, the war continued in Europe. Barometers of the Allies' alarm were the order books of the company. Between January and June, 1915, the Allies placed orders totaling $108,000,000 with Du Pont. Between July and September, they placed $137,000,000 in additional orders. By December, the Du Pont war contracts exceeded $340,-000,000, mainly for urgently needed smokeless powder.

Never before had the company been entrusted with a task of such magnitude. The tremendous powder orders were being filled, usually in advance of schedule. As the powder flowed out by shipload after shipload, a paralleling stream of earnings flowed back, its size in ratio to the enormous contracts involved.

The stock Alfred had thought too high at $160 a share in February was quoted at $390 in April, $670 in June, $775 in September. In the latter month, $30,000,000 in earnings had been plowed back permanently into the company's capital; two shares of new stock were issued for each share of old. A month later, the new stock was quoted at $430 a share! Even after writing off the sums expended on new factories, which would have no value after the war emergency, net earnings for 1915 were $57,800,000.

At this juncture, December, 1915, along with Francis I. du Pont and other stockholders, most of whom had comparatively small holdings, Alfred I. du Pont joined in a suit in the United States District Court of Delaware to compel Pierre and his associates to turn over to the company the stock they had bought from Coleman at $200 a share.

At the annual meeting of the stockholders in March, 1916, Alfred, Francis I. and William du Pont were not re-elected to the Board of Directors. Alfred was dropped as a vice-president.

The suit came to trial before Federal Judge J. Whitaker Thompson in June, 1916. Its basis was the contention that Pierre had withheld knowledge of pending war contracts at the time the stock purchase was under consideration by the Finance Committee, and, further, that Pierre and his associates had involved the company's credit by their syndicate's purchase.

Bankers, whose institutions participated in the loan to the syndicate, denied the loan had involved the company. Coleman du Pont testified briefly he had offered only 20,700 shares of stock to the company, that the offer to sell his entire holdings had been made only to Pierre, that he had sold at $200 a share, because he was satisfied with the price, although he fully expected the stock to sell higher if the war continued.

Almost a year passed before Judge Thompson rendered judgment. By questioning the propriety of some of the procedure followed by the syndicate, he upheld the plaintiffs' case in part. However, he left it to a vote of all the company's stockholders as to whether or not the syndicate's stock should be released to the company's treasury.

The syndicate's shares, amounting almost to one-third of the total, were barred by the court from being voted in this election. The result, reached in October, 1917, showed 312,587 of the eligible shares of stock voted in favor of Pierre as against 140,842 for Alfred and his co-complainants. Judge Thompson dismissed the case.

The plaintiffs appealed. The United States Circuit Court of Appeals found, in March, 1919, that Judge Thompson had been in error in ascribing any impropriety to Pierre and the syndicate. It held that Pierre had not abused his trust and that the company's credit had not been used. Dismissal of the suit was sustained.

Alfred's group attempted to appeal to the Supreme Court of the United States. They were denied. The case of Du Pont vs. Du Pont was closed, Pierre's management stood unchallenged, the distribution of Coleman's former holdings among the company's chief managers stood as made.

Alfred I. du Pont organized the Delaware Trust Company, for which he erected one of Wilmington's largest office buildings. Later he moved to Florida, where he became widely known in that state's banking and other affairs. Coleman du Pont founded his chain of hotels, laid the foundations in Delaware of one of the nation's finest highway systems, and in 1921 was appointed to the

United States Senate, the second Du Pont to serve his state in that body.

Both are now dead. Coleman passed away in 1930, Alfred in 1935. Perhaps Alfred du Pont's greatest service to the Du Pont Company was when, in 1902, he induced Coleman, and through him Pierre, to join him in taking over their common patrimony. He foresaw a much greater future for the company in explosives—and he was right. Coleman du Pont's vision merely began with explosives. It embraced not one industry, but industries and beyond.

The World War

Lord Moulton, Director-General of British Explosive Supplies, said in 1916 that the British and French armies could not have held out during the critical months of 1915 had it not been for the efforts of three American concerns—J. P. Morgan & Company, the Bethlehem Steel Corporation, and E. I. du Pont de Nemours & Company.

"The Du Pont Company is entitled to the credit of saving the British Army," said General Hedlam, chief of the British Munitions Board.

By the autumn of 1914, a wholly unforeseen situation, threatening an early and complete German victory, had developed in the conduct of the war.

The Allies had prepared for a war in the open. Consequently, they had laid up large supplies of light field guns and shrapnel for use against troops maneuvering in the field. Their factories were equipped to continue this supply in the quantities needed for open warfare on large scale. On this basis, by no means were the Allies unprepared for the conflict. However, they were unprepared for trench warfare, which actually came, while the Germans were far better equipped to take advantage of this development.

Trench warfare demands, not light guns and shrapnel, but heavy guns and high explosives shells.* A shrapnel shell is designed to burst with a scattering effect. Shrapnel is useful in destroying wire entanglements in front of trenches and in harassing troops in trenches or upon leaving them, but it is of no value in demolishing the trench system itself. On the other hand, the high explosives shell is designed for demolishing. Its mission is to penetrate a fortification and then explode with disruptive effect.

Long-range, big-caliber guns, with airplanes to direct their high explosives fire, plus machine guns in enormous numbers, composed the backbone of the German armament.

With trench warfare, too, the Allies faced a second unanticipated emergency. The old type of war, for which they had prepared, had been a series of separate, short battles, each of which was usually preceded by weeks of maneuvering for position. This new war was one of continuous battle! Day and night, week in and week out, each army pounded at the positions of the other. This unceasing siege went on along hundreds of miles of front. An ordinary day's routine of the entrenched battle-lines consumed more powder and shot than the Allies had contemplated for a large open engagement.

Climaxing this situation was the fact that the Germans held within their borders and in occupied territory the richest and most highly industrialized district of Europe—

*The high explosives employed in artillery shells should not be confused with industrial high explosives, such as dynamite, which cannot be used in cannon or in firearms of any kind.

best able to produce guns, shells and powder in steadily growing quantities.

In desperate and urgent necessity, Britain, France and Russia turned to the United States in the autumn of 1914. In American industry lay their one hope.

The first Du Pont war contract was closed with Russia on October 8th. It was for 960,000 pounds of trinitrotoluol, commonly known as T.N.T., for use in high explosives shells. Four days later, France ordered 8,000,000 pounds of cannon powder and 1,250,000 pounds of guncotton. Before the year ended, Allied orders placed with Du Pont aggregated 15,600,000 pounds of smokeless powder, mainly for cannon, 3,172,000 pounds of guncotton, and 2,160,000 pounds of T.N.T. During the following ten weeks these quantities were increased three and a half times. Before the end of May, 1915, Du Pont powder orders exceeded 107,000,000 pounds and T.N.T. orders, 22,000,-000 pounds. Also over this period 482,000 pounds of black shrapnel powder were ordered.

The emergency faced by the Du Pont Company in undertaking to fill these huge orders was as great, relatively, as that faced by the Allied governments. The company's total capacity for the manufacture of military smokeless powder in October, 1914, was only 8,400,000 pounds annually, or only a little more than the first French powder order. Within six months, it had contracted to supply almost thirteen times this original capacity, as well as large quantities of T.N.T. and guncotton, the latter chiefly for use in submarine mines and in torpedoes. Immediate expansion of plant facilities had been demanded on an unprecedented scale.

Smokeless powder had never been produced before in such amounts, or under emergency conditions. Expert knowledge was limited to a handful of men. Under the air-drying process followed in the Du Pont factories, which was in accord with the American military preference, it required sixty to ninety days to prepare finished powder for actual field use—an impossible delay under the circumstances. The stabilizing chemical, diphenylamine, was not available in this country, Belgium being the sole source of supply, other than Germany. Furthermore, the Du Pont nitrocellulose powder had to be fitted to English, French, Russian and later Italian guns, which had been designed each for its own type of powder. Du Pont had never produced smokeless powder except for guns of American design; a whole new series of powder sizes or granulations had to be worked out. Work formerly requiring months was to be done in days.

Thousands of workmen without experience in explosives manufacture had to be recruited, trained, and for the greater part housed. The United States was predominantly sympathetic to the Allied cause, but a strong pro-German sentiment also existed. The air was filled with stories of German spies and plots to blow up factories; powder factories made a shining mark.

Over all hung the uncertainty of the war's duration and of its ultimate result, not to mention uncertainty as to action by Congress to prohibit the export of war supplies.

With the first powder orders in October, such action began vigorously to be urged by friends of Germany in America. There existed, too, the certainty of general rises in wages and material prices should the Allied demands upon American industry and agriculture continue, in which event the hazardous nature of powder-making might make it extremely difficult to get and hold workmen, even at premium wages.

Against this background, the Du Pont powder plants—most of them as yet unbuilt or even on paper—became in 1914–15 the chief hope of augmenting the Allied powder supply. It was a responsibility wholly unanticipated and unintended by the nations whose future was at stake, and unforeseen and uninvited by the Du Ponts.

The company made clear, before a war order was signed, that it could not finance the sudden expansion of its plants demanded by the situation on the Allied fronts. A thirteenfold extension of facilities out of private resources was unthinkable. On its face, the task of effecting deliveries in time, in the quantities needed, appeared impossible, almost hopeless.

Allied agents offered to deposit securities in American banks to guarantee their payments after the delivery of powder and T.N.T.

"The United States is not in this war, gentlemen," replied Colonel Buckner, Du Pont vice-president in charge of military sales. "We can produce the explosives you need and we think we can produce them in time, but only if you assume the financial risks of an emergency expansion.

You are asking us to build costly plants that will have value only as scrap when the war ends. You are asking us to contract, on your behalf, for raw materials that may not be needed next month, or even next week, and which may never reach the stage of finished goods. I repeat, it's your war, and the risks must be yours."

The Allies accepted the risks.

Sums were deposited in the Du Pont treasury in the form of 50 per cent advance payments on powder and T.N.T. ordered at $1 per pound, a price high enough to cover the cost of building new plants. The organization assembled by Coleman and Pierre—the pick of the American explosives industry—went into action. Engineering and construction forces of 800 men were increased to an army that, at its peak, numbered 45,000. These men designed the machinery, put up the buildings, laid out and constructed the workmen's towns. While the plants were rising, the Experimental Station and Smokeless Powder Department dissolved the technical obstacles to large-scale production.

The sixty-to-ninety days' interval required for air-drying powder was reduced to eight days by adopting water drying. New powder types were worked out to fit the American propellant to Allied guns. A process for the manufacture of diphenylamine from coal tar was perfected, a plant erected at Repauno. In 1916, for the first time, that vital chemical was being produced in quantity on American soil.

At Hopewell, on the James River in Virginia, the

world's largest guncotton plant mushroomed, to give employment to 28,000 persons and supply thousands of tons of nitrated cellulose. A police force of 1,400 uniformed men was organized under Major Richard Sylvester, formerly Commissioner of the Washington, D. C., Metropolitan Police. Its mission was to maintain order and to see that rumors of trouble were not translated into facts.

From a starting capacity of 8,400,000 pounds a year in October, 1914, Du Pont smokeless powder plants located at Carney's Point, Haskell and Parlin, in New Jersey, attained a capacity rate of 200,000,000 pounds a year before the end of 1915. In December, 1916, the capacity reached 290,000,000 pounds a year. When the United States entered the war in April, 1917, the three plants were capable of supplying powder to the Allies at a rate of 357,000,000 pounds a year. Just twelve months later —April, 1918—their capacity had reached 455,000,000 pounds, or fifty-four times that of October, 1914.

Concurrently, the production of T.N.T., which in 1914 had amounted to 660,000 pounds monthly and chiefly was of a grade suitable for industrial use only, was increased almost to ten times that amount and of the higher purity demanded for artillery shells. The company had had no prior experience in the manufacture of such military high explosives as tetryl, picric acid and ammonium picrate, but it was producing 435,000 pounds of these monthly by 1917.

All told, in four years of war, Du Pont supplied the Allied forces with almost one and one-half billion pounds of military explosives, of which the powders for propelling shells amounted to 40 per cent of the total amount of such powder consumed. In addition, it supplied American industry for the production of coal, metals, cement, building stone and the like, with more than 840,000,000 pounds of dynamite and black blasting powder, which was one-half of the nation's total requirements for the war period. The annual capacity of all Du Pont plants, before the end of the war, stood at 893,000,000 pounds of explosives.

To accomplish this work, the company employed at the peak more than 100,000 persons. It built, in addition to its war plants, 10,790 workmen's homes, which, with hotels, boarding-houses, women's dormitories and bunk-houses, were capable of housing 65,000 persons. Power-houses required the continuous development of 200,000 boiler horse-power. Pumping-stations had a capacity of 305,000,000 gallons a day, greater than the combined daily water supply of the cities of Philadelphia and Boston. Refrigeration apparatus produced the equivalent of 9,350,-000 pounds of ice daily, a consumption equal to that of the city of Chicago; 10,700 tons of coal were burned every twenty-four hours.

The engineers built 150 miles of guard fences, 100 miles of standard-gauge and 208 miles of narrow-gauge rail-roads. Car classification yards could handle 1,600 cars at one time, an essential factor in dealing with the enormous quantities of materials in ceaseless movement in and out of plants. These included a grand total of 2,660,000 bales of cotton linters, 1,400,000 tons of nitrate of soda, 460,000 tons of sulphur, and 216,500,000 gallons of alcohol, to mention only a few major items. From these basic materials the plants produced almost two billion pounds of nitric acid, two and a half billion pounds of sulphuric acid, and 1,158,000,000 pounds of nitrocellulose.

On experimental work, research, and chemical control, a thousand men were employed and the company's laboratories expended $3,360,000.

Inevitably, there were accidents. Tens of thousands of new workmen could not be absorbed into explosives plants operating under war conditions without entailing risks. Explosions in four years destroyed $6,700,000 in property and cost the lives of 347 men. The accident ratio per thousand of employees, however, was actually less than it had been in the years previous to the war, astonishing evidence of the efficiency of the company's safety division and the plant guards.

Throughout the war, not a single lot of Du Pont-made smokeless powder was delivered behind schedule time, not one pound was returned as unsatisfactory.

VI

Old Hickory

WITH the breaking of diplomatic relations with Germany on February 3rd, 1917, it became apparent that the United States could not stay out of the war much longer. To Du Pont officials, it was equally obvious that American participation would necessitate an early increase in the country's powder supply. Practically the entire Du Pont output was being consumed by the Allies. The nation's additional facilities, made up of two comparatively small Government-owned plants and those of three private companies, one of which was bankrupt, were insignificant as compared with the possible need.

Du Pont engineers were sent at once into ten states to search for new plant sites within safety zones prescribed by the War Department. The preferred locations chosen were one near Charleston, West Virginia, a second near Nashville, Tennessee, and a third near Louisville, Kentucky. America's declaration of war on April 6th, 1917, found the company not only ready to proceed with any construction program the Government might direct, but urging that such work be inaugurated without delay to provide for the almost certain needs of 1918.

A meeting of the four powder manufacturers was called

by the Army Chief of Ordnance on April 20th.* The Army estimated its smokeless powder requirements over the next twelve months at 78,500,000 pounds. Du Pont offered to furnish the entire amount at 60 cents a pound for water-dried rifle powder; 47½ cents for water-dried cannon powder; and 50 cents for air-dried cannon powder. Since 1913, when Congress had fixed the cost of air-dried cannon powder at 53 cents, the costs of raw materials, labor and other items had advanced 18 cents per pound of powder, so that, on the basis of costs, the prices quoted were 30 per cent below the 1913 standard of value.†

At this meeting "the companies were informed that it is not the present policy to urge an increase of capacity where existing free capacity is in excess of estimated requirements."

However, no excess capacity existed. The Du Pont excess, so-called, was simply the result of failing to renew Allied contracts as they expired, for the purpose of diverting powder to the United States. Again on July 30th, 1917, in a letter to F. A. Scott, Chairman of The General Munitions Board of the Council of National Defense, the company directed attention to the danger of a serious powder shortage. Colonel Buckner wrote:

The Ordnance Department of the Army has already asked us to reserve for the Army our entire unsold capacity, and we

---

*All facts set forth in this chapter are based on documents and evidence submitted before the United States Senate Special Committee Investigating the Munitions Industry, 1934–35.

†Water-dried cannon powder, offered at this time at 47½ cents a pound, was to constitute the bulk of the Government's needs.

feel confident that the Allied Governments will require large additional quantities of powder. In fact, we understand informally that the United States Government will require within the next twelve to fourteen months close to three hundred million pounds in addition to the capacity we have available within that time. How this large amount of powder is going to be supplied to the United States and the Allied Governments we do not know, unless all the Allied Governments confer upon the subject and determine upon a plan of procedure which will involve the construction of increased capacity. . . . Owing to the increase in price of various materials—labor, etc.—this expense may be considerably increased if there is any delay in starting this work. . . .

One week later, the company was forced to reject outright an order for 20,000,000 pounds of powder needed by Russia, which, torn by revolution, the Allies were trying desperately to hold in the war. From France, General Pershing appealed for one million American troops by the following May. As to powder, "the situation in France was such that its factories working at their utmost capacity fell short nearly 19,000,000 pounds per month of the quantity required by the French Armies."‡

Still, nothing was done until October. Then, suddenly, the official estimate of powder needs was revised. The War Industries Board "determined that the total estimated requirements of powder for the Allies and the United States to be produced in the United States up to November 1st, 1918, was 1,016,748,000 pounds, and that the capacity of the plants in this country was about 472,000,000 pounds of powder"—in other words, that American factories could produce less than half of the powder needed. The board

‡Official memorandum of the War Department.

voted unanimously "that immediate and drastic action should be instituted, as the explosives situation is regarded as critical and of supreme importance."

The Du Pont Company, on October 3rd, 1917, was asked to submit immediately a proposal to build and operate powder plants of a million pounds powder capacity per twenty-four-hour day, the work to be on a "cost-plus" basis. Five days thereafter, on October 8th, the Du Pont proposition was in the hands of Major-General William Crozier, Army Chief of Ordnance. It estimated construction costs at $90,000,000. Operating costs for twelve months, the period stipulated, would approximate an additional $180,000,000. The project was the largest in the history of the War Department.

Terms and other details were agreed upon in less than three weeks and on October 25th, 1917, General Crozier issued a formal order to the Du Pont Company to proceed with the preparation of all plans and the optioning of necessary land, subject to the anticipated approval of the order by the Secretary of War. The order provided for two huge operations to be carried out on the best of the three sites selected earlier by the company's engineers. In all, there were to be ten producing units, each with a capacity of 100,000 pounds daily. The first unit was to come into operation eight months after purchase of the land; an additional unit was to be completed every thirty days thereafter. A maximum time of eighteen months was allowed for consummation of all building.

For the performance of this work, as the Government's

agent, the Du Pont Engineering Company was to be formed. Its officers were to be those of the Du Pont Company proper. Its responsibility was to be guaranteed by the parent concern. Further, the order provided:

The Government will pay directly or will reimburse you for all costs of the construction of the plants, and in addition you will be paid at the time of making payments on account of such construction, whether directly to you or in reimbursement—

(a) A sum equal to seven (7) per cent thereof to cover preparation of plans, the procurement of sites, engineering supervision other than local supervision, and services in connection with purchasing and forwarding deliveries of materials necessary for the construction of the plants, and

(b) A sum equal to eight (8) per cent to cover administration other than local administration, pro rata share of overhead expense other than local overhead expense, and to cover profit.

For powder produced and accepted, the Government agreed to pay all operating costs plus a fee of 5 cents per pound above an estimated base cost of 44½ cents per pound, regardless of the size of granulations. Any reduction in the base cost effected through manufacturing efficiencies or inventions was to be divided evenly by the company, as operator, and by the Government, as owner.

General Crozier and the Army Ordnance experts considered these terms most favorable to the United States. The estimated cost plus operating fee, which made the base price 49½ cents, was equivalent to the price of 47½ cents quoted earlier by the company for powder from its own plants, plus 2 cents a pound approved by the Ordnance Department to allow for a subsequent increase in alcohol prices. Neither quotation included any charge for amortization of plants. Moreover, the price of 49½ cents a pound, the experts held, was lower than the cost for which the Army could produce powder at its own factory at the Picatinny Arsenal.

Du Pont engineers swung into their biggest assignment of the war on October 25th, 1917. They had broken all records for the Allies. Now they were asked to better those records for the United States, a task that was to be superimposed on manufacturing requirements already tremendous. Overnight, the company's executives had had their responsibilities and supervisory duties doubled.

In less than a week, land was being placed under option near Charleston, West Virginia, regarded as the most favorable site, because it was served directly by a railroad, whereas the second-choice site in Tennessee was seven and one-half miles from the nearest rail line. With equal dispatch, contracts were made ready for orders of $2,369,000 worth of mechanical apparatus. Wilmington machine shops were bought outright to augment the company's own shops.

Hundreds of draftsmen, working in continuous shifts, began revising sixty acres of Du Pont blueprints, the accumulation of the war, to make them adaptable to the specific requirements of the Charleston project. In addition to these revisions, work was begun on about forty acres of new drawings requiring completely new designs.‖

‖These powder plants were of such magnitude and detail that the drawings required could be measured in acres.

Once again went out a Du Pont call for men, this time in competition with recruiting sergeants, draft boards, shipyards, farms, with all American industry seething in the emergency of a nation facing the world's greatest war crisis.

This mightiest of powder-making operations was surging toward reality the morning of October 31st, 1917. Hourly, the preparations of the engineering machine that had been delegated to build it were gaining in power and momentum. Then, abruptly, all work was stopped. The following telegram had been handed to Pierre du Pont:

Have just had presented to me the details of the proposed contract with regard to increased capacity for powder production. The matter is large, intricate and important. Do nothing until you hear further from me. Stay all action under the order until I can acquaint myself thoroughly with all features of the matter.

NEWTON D. BAKER, SECRETARY OF WAR

Probably the chief reason for Secretary Baker's telegram was that while the War Industries Board had been informed orally on at least two occasions of General Crozier's negotiations with the Du Pont Company, the board had not been given an opportunity to pass finally on the order before it was signed. Thereafter, apparently, Secretary Baker was guided largely by an investigation made by Robert S. Brookings, a member of the board, whose inquiry into the powder business was limited to an afternoon's visit on November 6th to the naval powder plant at Indian Head, Maryland.

Brookings at once wrote Secretary Baker that the

Du Ponts would realize a net profit of 15 per cent or $13,-500,000 on the construction authorized, and possibly $30,-000,000 on operation of the completed plants. He continued:

From statements made to me by Mr. Patterson, superintendent at Indian Head, I would place very little value upon the super-expert knowledge of the Du Ponts, or patents or other privileges attached to their services. They probably produce at Indian Head everything provided for at the proposed new plant except possibly a small amount of diphenylamine.

I herewith hand you enclosures just received from Col. Pierce, which indicate the cost to the British Government for contractor's services on a plant probably similar to this would average about 2 per cent. This does not seem, however, to cover the plans and specifications. . . .

That the price demanded by the Du Ponts for construction, service and operation is utterly out of scale with any possible service they can render would seem obvious.

This letter, sent to Secretary Baker by special messenger, precipitated a controversy that was to hold up work by the Du Pont Engineering Company for three months, at a time when much of the Army's plan of action hinged on the immediate provision of an adequate powder supply. Lieutenant-Colonel C. T. Harris, Jr.,§ Army Ordnance expert, later testified:

"The time element was the important thing. . . . General Crozier thought he was making an equitable contract. . . . There were three months lost in the middle of the war by these negotiations. That had a very serious effect

§In 1942, Major-General and Assistant to the Chief of the Ordnance Department.

on the military effort. Fortunately it did not have a fatal effect, but it might have had."

Indignantly General Crozier insisted the proposed contract not only was fair but that it "evidenced the greatest generosity upon the part of the Du Ponts." He told the War Industries Board that, in his judgment, which was backed by every expert of his staff, "the emergency necessity overshadowed all question of cost." In his report of that session to Secretary Baker, Brookings wrote that the Ordnance Bureau's experts had recommended the terms set forth in the Crozier order "in the most unqualified way" and that "no amount of arguments made any impression on them." He added, "You can readily see how difficult it is to get men who have so conclusively passed upon a proposition to admit that they have made a mistake."

At this juncture, the Du Pont Company volunteered to relinquish the order to any agency the Government named, if by so doing the country's needs could be more speedily attended. At the same time, the company submitted a detailed estimate of direct charges payable out of commission that, it said, precluded any possibility of profit beyond 7½ per cent and foreshadowed much less, not excepting the possibility of actual loss. Considering indirect charges too, the likely profit was estimated by the company at about 3½ per cent.

Brookings' comment to Irénée du Pont was that he would rather have the United States pay $1 a pound for powder, sold without profit, than to pay 50 cents a pound if that price allowed 10 cents of profit to the Du Ponts.

Brushing aside the advice of General Crozier and other Army Ordnance officials, the Secretary of War moved to enlist the services of a large construction company to build the needed plant under governmental direction. The War Industries Board decided, on November 26th, 1917, that the construction company lacked the "technical expert service" the work demanded. The board's chairman, Daniel Willard, wrote:

"We are further of the opinion that the emergency nature of the case requires that the proposed plant be constructed in the least possible time and operated with the greatest efficiency, and based upon the best evidence we have—which is that of demonstration—it is our opinion that the Du Pont people are in every way the best fitted for securing this result."

Willard suggested that $1,000,000 be paid the Du Pont Company on account, and that the balance of its commission should be fixed by arbitration on completion of the work. The company rejected this plan, finally proposing the matter be submitted to impartial arbitrators.

Secretary Baker replied, December 12th, that he considered the matter closed with the company's refusal to accept the Willard proposal, that "the question would not be reopened," and that he had already "proceeded to work out a plan for the direct creation of this capacity by the Government itself." He appointed D. C. Jackling, of San Francisco, special agent of the War Department, to build and operate all necessary powder plants, with authority to proceed according to his best judgment.

Jackling, an outstanding executive of the copper-mining industry, was without experience in powder manufacture. One of his first moves was to try to secure the services of the Du Pont chief engineer. Major Ramsay had died in 1916 and Harry M. Pierce now occupied that post—the same Harry Pierce of early Repauno days, who had quit his job to spend four years in engineering school.

"You can't do it that way, Mr. Jackling," said Pierce. "You need an organization of trained men, not just one man. This is a case of national emergency. The work must be done as quickly as possible. In my opinion, there is only one way to handle it without costly delays, and that is, let the Du Pont engineering organization undertake the job with the backing of the whole Du Pont Company. I wouldn't attempt such a gigantic task under any other conditions."

However, the Du Pont Company volunteered help. Almost 60 per cent of the drawings for the Charleston plant were completed. The company offered to turn these over to Jackling at no cost to him. It agreed to lend him drawings of existing Du Pont plants for use as references in getting up the remaining drawings for Charleston. The Du Pont plants were to be opened to Jackling's engineers for study. Du Pont engineers would go to Charleston and lay out the plant there on the site chosen.

Thus assured, Jackling engaged the Thompson-Starret Company, nationally known construction concern, to build a powder plant of 500,000 pounds daily capacity on the Charleston site, which was now officially designated as "Nitro." Thompson-Starret were specialists in office buildings, loft buildings, and power plants.

The need, however, was for one million pounds daily output. Convinced after a swift inquiry that the Du Ponts would have to provide the additional 500,000 pounds capacity, because no other organization was equipped for the task, Jackling reopened negotiations with the Du Pont Company with a view to building on the less favorable site near Nashville, to which was given the name of "Old Hickory." The Secretary of War authorized this action by Jackling.

In the earlier dispute over what was overhead and what was profit, the company had estimated its likely actual profit under the Crozier order at $3\frac{1}{2}$ per cent of construction cost, and it had also suggested the possibility of loss.

"Did you mean that, Mr. du Pont?" asked Jackling.

"We did," said Pierre. "Prepare a contract that will assure us 3 per cent net on construction, $3\frac{1}{2}$ cents a pound on operation, one-half the saving we might effect below an agreed base cost, and a reasonable sum, say $500,000, to cover the cost of designing and preparing plans, and we'll consider those terms better than those the Secretary of War refused to approve."

Such a contract was drawn. It was signed by the company on January 29th, 1918. In all its essential features, it was the Crozier contract of October 25th, 1917, so far as profit to the Du Ponts was concerned. During the three months of delay, however, the index of construction costs published by the *Engineering News-Record* had advanced from 167.1 to 184.5. The cost of the Old Hickory plant, including the necessary railroad connection, was placed at $50,000,000.

The delay, moreover, had aroused the military authorities to a state of serious alarm. If the Army was to obtain powder in the quantity needed in time, they said, a third 500,000 pounds plant would have to be built immediately. Jackling pointed out that no organization existed which was capable of building a third plant, that the only solution was to augment the capacity of the Old Hickory and Nitro operations. So precarious was the situation regarded in Europe, with the German Western front being steadily reinforced from the East, that nobody dared even to guess the outcome of the approaching spring and summer.

The time was now late February, 1918. Jackling conferred with Du Pont representatives. Could the Du Ponts build a 900,000 pounds plant at Old Hickory within the time limit of their contract? The answer was "Yes, if you let us do the job in our own way!" Under the then existing arrangement, it was necessary to submit for approval all plans and proposed expenditures to Jackling's office in New York before going ahead with any work. This was a cause of constant worry, delay and endless explanations of technicalities.

Pierre du Pont wrote:

You have explained to me the urgency of this work, as well as the necessity of its undertaking by E. I. du Pont de Nemours & Company. It is only this necessity that warrants our considering the burden. The latter is so great that the question of compensation is of less importance than the question of the saving of man-power, especially that of the administrative end of the business.

I therefore urge upon you to arrange that the carrying out of this work may be placed in the hands of our company entirely without hindrance as to supervision and approvals. In return for this concession, we are prepared to place the question of compensation beyond possibility of criticism.

A new contract was executed on March 23rd, 1918. It provided for a plant of nine units, each of which was to be of 100,000 pounds daily capacity. The estimated cost was first placed at $75,000,000, but rapidly rising costs soon made it necessary to increase this figure to $90,000,000. Jackling stepped out and the Du Pont Engineering Company was given free rein, subject only to verification of its accounts by Army auditors on the ground. The Du Pont construction fee was fixed at One Dollar, and the company voluntarily waived all profit that had accrued to it under the Jackling contract made earlier. The profit for operation of the plant remained the same.

Even Du Pont critics have conceded that the speed with which this gigantic plant was built and put into operation stands as the greatest engineering feat of the war. On July 2nd, 1918, three months and nine days after the contract of March 23rd was signed, the first 100,000-pound powder unit at Old Hickory was finished and started in production. This was forty-four days ahead of the March contract schedule and eighty-one days ahead of the January contract schedule. It was almost precisely the date set for completion of the first unit under the Crozier order of October, 1917! Du Pont men had made up the time lost.

The second 100,000-pound unit was completed twenty-two days ahead of the March schedule; the third, fifty days ahead; the fourth, fifty-three days ahead; the fifth, eighty-two days ahead; the sixth, sixty-six days ahead. When the war ended on November 11th, the aggregate saving of time on these six units alone was 317 days, and Old Hickory as a whole was 93 per cent complete!

The January contract had not anticipated the completion of five units before March 16th, 1919, or more than four months later!

Consider the accomplishments at Old Hickory. Up from vacant fields, miles from a railroad or from any community of size, sprang a fully organized city of 30,000 persons. It contained 3,867 buildings—homes, apartments, hotels, restaurants, schools, churches, theaters, hospitals, a city hall, fire and police stations. Streets were paved. Sanitation was modern. Water was filtered. In the course of ten months, approximately 250,000 persons were employed on the work for shorter or longer periods.

The powder plant operation itself called for the erection of 1,112 buildings of every size and style of construction known to modern building practice. All but 100 of these structures were completed, along with 99 per cent of the intricate mechanical equipment essential for the processing of raw materials from the basic state through to finished gunpowder.

The task was replete with exploits in engineering. Seven

and a half miles of single-track railroad, including a trestle of piling nearly one-half mile long with 400 feet of fill averaging 20 feet deep, were built in thirty days. This spur, connecting with the nearest existing line, soon had to be double-tracked to handle at its peak 1,100 cars and 31,000 passengers daily. A steel suspension bridge 540 feet long and with 1,260 feet of trestled approaches was thrown across the Cumberland River in less than five months after high water had repeatedly wrecked a pontoon bridge.

The rate of construction expenditure for the whole operation over ten months was equivalent to two and one-half times the maximum rate of expenditure for any year on the Panama Canal. Total cost was $85,000,000.

With the coming of peace, Major-General C. C. Williams, successor to General Crozier as Chief of Ordnance, wrote the Du Pont Company:

> To have built up and put into operation the first units of such a plant in less than five months from the date of breaking ground, under the stressful conditions existing, involving as it did the construction not only of the major plant but of a number of sub-process plants, each of which in themselves might be regarded as an undertaking of no little magnitude, must always be regarded as a remarkable achievement. . . .
>
> The history of what you have done at Nashville is paralleled by the very satisfactory and uniform fulfillment of your expectations on practically all other work you have undertaken for the Government. All of this on materials that were most vital to the successful issue of the war.

War Profits

THE sudden ending of the war brought an almost immediate cessation of Du Pont war activities. The number of munition plant workers was reduced from 85,600 to 18,000 in seven weeks. The latter were held, temporarily, on orders from Washington as a precaution against possible renewal of hostilities.

Du Pont war contracts suspended totaled approximately $260,000,000. Upwards of $10,000,000 was saved the United States by halting the delivery of any further materials, and an additional $4,000,000 was saved it by the company voluntarily closing down operations without waiting for suspension orders from Washington.

All told Du Pont companies had more than 100 war contracts in effect at the Armistice. Many of the contracts had been imposed because the already overworked Du Pont technical staff was the only resort in the emergency. For example, a shortage of toluol had made it mandatory to find a substitute for T.N.T. The company's chemists had answered with T.N.X., or trinitroxylol; while the toluol deficiency later had been remedied by the manufacture of that chemical in quantity from coal-tar derivatives and gas house oils. Need had arisen for a new explosive to load airplane drop bombs, one as powerful as

T.N.T., sensitive enough to be readily detonated, but insensitive enough to withstand piercing by bullets in aerial combat. This need had been met with 1,200,000 pounds of "Lyconite," made at Repauno.

On demand, Du Pont engineers had furnished 87,000 demolition outfits for field troops and 8,000 smoke boxes to screen American ships from submarines. Chemists had developed new powder containers for trench mortars and gas projectors, had devised fire-and-gas-resistant fabrics for dugout curtains and clothing, pyroxylin dope for airplane wings, unbreakable eye-pieces for gas masks. The Du Pont Company and Du Pont American Industries, a company formed especially for the purpose, had acted as the Government's agent in the purchase, storage and distribution of 683,000 bales of cotton linters.

The Old Hickory plant had been only one of five major contracts assigned to the Du Pont Engineering Company. The latter had also built and was operating a shell-loading plant at Penniman, Virginia, it had under way a high explosives plant at Racine, Wisconsin, and it was operating plants for the bag-loading of cannon powder at Tullytown, Pennsylvania, and at Seven Pines, Virginia, all vitally important projects.

Had the war continued another year, the United States would have produced more than one billion pounds of smokeless powder in 1919; three-fourths of this would have been from Du Pont operations.

Scarcely had Armistice Day passed before rumors of scandals began to be heard. The size and spectacular nature of Old Hickory made it a target. In 1919, the rumors became the subject of official attention. Their burden was that the Du Ponts had cheated the United States out of enormous sums. The charges ranged from petty thievery to wholesale criminality and flagrant fraud.

If these charges had been confined to the immediate post-war period, and if the exhaustive investigations that established their falsity had also silenced them, they could be passed over here as incidents of little consequence. However, time and again the charges have been revived, each repetition being made with the air of discovery. In a speech delivered July 14th, 1922, from the floor of the United States Senate, a member of that body declared:

I have here a tabulation drawn from official sources showing that, beginning with January 10th, 1918, and running through to a latest date of February 11th, 1919, the Du Pont Company and two of its subsidiaries secured advances from the Federal Treasury in the gross sum of $99,250,552.80. This money, Mr. President, was paid for the purpose of creating huge establishments which bear the Du Pont name and which were devoted to furnishing supplies to the Government during the war. Thus the people of the United States not only financed the Du Ponts in the tremendous extension of their business but were also mulcted by the Du Ponts to an extent which enabled the company to increase its plant value to an admitted $220,000,000 while at the same time taking out net profits which in one year amounted to $129,000,000.

One year and a half later, a Tennessee newspaper reported "detailed charges of wholesale fraud amounting to between $50,000,000 and $60,000,000 in the erection and operation of the Old Hickory powder plant at Nashville during the war." Again, authority was given the statements by the fact that the newspaper's quoted source of information was the United States District Attorney at Nashville. Said the newspaper:

These alleged frauds comprise only a few of those as charged by the Government in the various reports of the expert auditors who have been working on the investigation since 1919, when the matter was first brought to the attention of the Department of Justice. Many of those employed at the plant at that time, and who have made affidavits and statements since the investigation was begun, differ as to the amount of recovery that could be obtained by the Government through a full investigation of the entire construction and operation of the plant. Some place the figures merely among the millions, while others estimated that as much as $100,000,000 could be recovered from the alleged frauds.

The "frauds" enumerated included the dumping of excess war supplies upon the Government, "wholesale faking" of freight vouchers, gross overcharges in the purchase of materials, gross irregularities in the payment of labor, even down to the cheating of dead victims of the influenza epidemic out of burials and theft of their personal belongings.

In the meantime, the Government had assigned auditors to investigate all dealings of the Du Pont Engineering Company with the United States. For three years they had been at work. Simultaneously, the Department of Justice had employed its own investigators. Special attorneys were named by the Attorney-General to delve into the legal issues as they were raised.

At one time, 103 auditors were at work in the Du Pont Wilmington offices alone. The expense, as of May, 1922, stood at "approximately $1,000,000." The hunt continued. The floorboards in long-vacated offices at Old Hickory were torn up in the search for records suspected of having been hidden. The refuse dumps were raked. Ex-clerks were questioned. So intense was the effort that some vouchers were worn into illegibility by the number of Government men who handled them.

The result, announced on August 5th, 1925, after six and a half years of auditing, checking, probing, and general inquisition, including numerous official hearings, was summarized by Attorney-General John G. Sargent in a communication to the Secretary of War, thus:

> An extensive investigation of the charges of fraud or crime above referred to has failed to disclose reasonable or probable grounds for believing that during the performance of the contracts in question the Du Pont Company committed the offenses with which it has been charged, or any crimes, and in my opinion there is no warrant either for further investigation along these lines or for the institution of proceedings against the company based upon charges of fraud or crime.

The "as much as $100,000,000" that the anonymous investigators of the Tennessee newspaper's story estimated "could be recovered from the alleged frauds" dwindled ultimately to about $300,000, which was the total of all disallowances made by the Government out of Du Pont Engineering Company expenditures on contracts aggregating $129,000,000.

The auditors conceded that the disputed $300,000 had actually been spent on behalf of the United States. About one-fifth of it was chargeable to ordinary clerical errors, and the remainder was the result of disagreements over legal technicalities in the contracts. These disallowances, amounting to one-fifth of one per cent of the total expended and which in no way involved fraud, the company accepted after seven years of harassment in order to close the account.

As to the "enormous profits" supposedly realized by the Du Pont Engineering Company as an outgrowth of its transactions with the United States at Old Hickory and elsewhere, the final settlement of 1925 was likewise revealing. During the two years of 1918 and 1919, as per Government agent's report, the engineering company's gross profits on war work had aggregated $2,451,185.88. Income and profits taxes paid against this gross had totaled $1,976,645.45. Bonus awards to operating employees, which were borne entirely by the parent company, amounted to $193,194.18. The net profit realized by the Du Pont Engineering Company was $281,346.25.

Seemingly the investigation, the Attorney-General's opinion, and the settlement of October, 1925, would have answered the allegations for all time. The Tennessee newspaper, convinced of the falsity of its report, published a full retraction and expressed its regret to the Du Pont Company. Over the signature of Hanford MacNider, Acting Secretary of War, the Government declared that the settlement "shall constitute full and complete satisfaction and termination of any and all claims and demands in law or in equity by or on behalf of the United States against the contractor, and/or by the contractor against the United States, in connection with each and all of the contracts aforesaid."

However, in 1934, the Old Hickory contract was dug up once again by the Senate Committee Investigating the Munitions Industry, and once again the laid ghosts were paraded. Two volumes of testimony and exhibits, on Old Hickory alone, were published at the public's expense. Musty records were reprinted. Old witnesses were recalled. One brief statement by Major Arthur Carnduff, summoned to the stand, is illustrative of the result.

Major Carnduff had been one of the special legal assistants to the Attorney-General in the earlier investigation. In fact, he had been responsible for the case as the Department of Justice attorney immediately in charge. He was asked by Alger Hiss, the Senate Committee's examiner, if, had he drafted his own opinion, he would have agreed with the opinion written subsequently by Attorney-General Sargent exonerating the Du Pont Company. The record reads:

MR. CARNDUFF. Regarding fraud and crime?
MR. HISS. Yes.
MR. CARNDUFF. At the time I left the case, I would. After a year's investigation and the reports of various auditors, detectives, and investigators, at the time I was taken off the case I had no positive evidence whatsoever of any fraud or crime on the part of the Du Pont Company or any of its subsidiaries, and I had nothing to lead me to suspect that there was fraud or crime.

Two more facts should be stated from the record:

(1) Not one dollar of money advanced by the United States went into the expansion of the Du Pont Company's own plants, either during the war or after it. The Government plants in entirety were turned over to the Government early in 1919. Five years later, however, the company bought from the Nashville Industrial Corporation (in which Du Pont had no interest) the town of Old Hickory and a tract of land for use in strictly commercial operations in no way related to war or to explosives, or for which none of the old war equipment and buildings were adaptable. The price paid the Nashville company was upwards of $800,000. Consequently, Du Pont factories built wholly from private funds now stand on the site associated with this old war plant.*

(2) Study of the record should convince any impartial investigator that the action of Secretary of War Baker in suspending the Crozier order of October 25th, 1917, under which the Du Pont Company would have begun forthwith the building of new powder manufacturing plant, resulted ultimately in a large needless extra cost to the nation. The Crozier order was believed fair by every responsible officer of the Ordnance Bureau. It provided for an additional powder-plant capacity of 1,000,000 pounds per 24-hour day, which was supposedly sufficient to meet all the augmented requirements of the Government. How-

*According to testimony before the Munitions Committee, the Government obtained approximately $3,500,000 in the sale of land, buildings and other materials at Old Hickory, of which the property acquired by Du Pont was only a part.

ever, the alarm caused by the delay which followed the Secretary's action, and the fear that the Army's needs could not be filled before those needs became dangerously acute, led the Government to contract for a new capacity aggregating 1,550,000 pounds per day, and at increased construction costs. . . .

The story of Du Pont powder prices for 1914–18 is probably unique in warfare. The Congress that, in 1913, had fixed the price of smokeless cannon powder at 53 cents a pound and of smokeless small-arms powders at 75 cents a pound, had been still under the influence of the sensationalized charges of a "Powder Trust" lawsuit. Most European governments were paying more for their powder. The United States' prices were exclusive of the cost of the air-tight, zinc-lined containers that were provided, necessarily, for all powder shipped abroad at an additional cost of 5 cents to 6 cents per pound.

When the war started, Du Pont was selling no powder in Europe. Because of obvious risks, the company moved most cautiously in accepting orders from the Allied Nations. The contracts entered into at the price of $1 per pound were strictly commercial transactions at a time when America had no part in the war. The price was agreed to include provision for the cost of plants that would be useless with the coming of peace, then expectedly not distant. It anticipated, too, a rapid rise in all manufacturing costs, accidents and other trouble.

All costs did rise. Accidents did occur. Yet within two

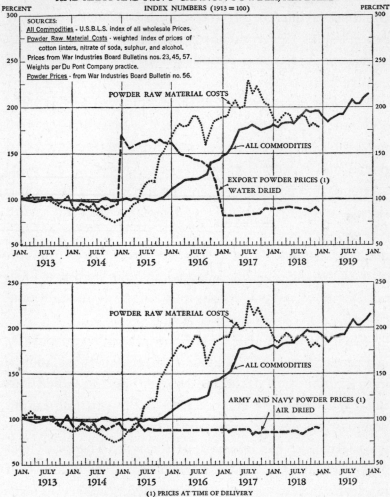

PRICE PER POUND OF EXPORT CANNON POWDER, WATER DRIED
AND ARMY AND NAVY CANNON POWDER, AIR DRIED

PERCENT   INDEX NUMBERS (1913 = 100)   PERCENT

SOURCES:
All Commodities - U.S.B.L.S. index of all wholesale Prices.
Powder Raw Material Costs - weighted index of prices of
    cotton linters, nitrate of soda, sulphur, and alcohol.
Prices from War Industries Board Bulletins nos. 23, 45, 57.
Weights per Du Pont Company practice.
Powder Prices - from War Industries Board Bulletin no. 56.

POWDER RAW MATERIAL COSTS

ALL COMMODITIES

EXPORT POWDER PRICES (1)
WATER DRIED

JAN. JULY JAN. JULY JAN. JULY JAN. JULY JAN. JULY JAN. JULY JAN. JULY JAN.
    1913   1914   1915   1916   1917   1918   1919

POWDER RAW MATERIAL COSTS

ALL COMMODITIES

ARMY AND NAVY POWDER PRICES (1)
AIR DRIED

JAN. JULY JAN. JULY JAN. JULY JAN. JULY JAN. JULY JAN. JULY JAN. JULY JAN.
    1913   1914   1915   1916   1917   1918   1919

(1) PRICES AT TIME OF DELIVERY

The prices received by the Du Pont Company for its powder were if anything slightly lower than those indicated by these charts, because when the United States entered the war some of the other American manufacturers of smokeless powder were able to show the Government that they could not meet the Du Pont Company's costs, and on that account the Government paid them higher prices than the Du Pont Company was charging.

years the Allies were paying no more for Du Pont powder, water-dried, than the United States' prices of 1913 for air-dried powder, plus the cost of boxes!† On cannon powder, which comprised the bulk of the demand, the company's voluntary reductions in price per pound were:

To 97½ cents in April, 1915
To 80 cents in November, 1915
To 57½ cents in July, 1916
To 55 cents in September, 1916

When the United States entered the war in April, 1917, and thereafter until the Armistice, the Du Pont prices to the Allies, after allowance for boxes, were identically those charged the United States. The prices of small-arms powder, T.N.T., and other military explosives, underwent comparable reductions.

Considerably more than one-half of the Du Pont war-time powder output was sold at 1913 prices or less. At the war's end, the company's prices were from 6 per cent to 20 per cent lower than in 1913, although raw materials were costing 132 per cent more than in 1913. Meanwhile, between June, 1914, and June, 1918, No. 2 hard winter wheat at Kansas City rose from 85 cents to $2.12 a bushel, or 149 per cent; Minnesota flour advanced from $4.65 to $12.00 a barrel, or 158 per cent; middling cotton

---

†Most of the powder made during the war was water-dried, whereas powder supplied the United States Government prior to war was air-dried. The cost of air-drying is slightly (about 2½ cents a pound) higher than water-drying, for which allowance should be made in the figures that are given.

prices at New York City increased 125 per cent, print cloth 300 per cent, steel 151 per cent, while the United States Bureau of Labor price index for all commodities advanced 91 per cent.

Powder prices alone came down during World War I!

BOOK FOUR

The Plowshares

★ CHAPTER ★

I

The Magical Age

DESPITE the World War and its gigantic divergency, the Du Pont Company continued to push its program of expansion "beyond explosives" into the fertile fields of chemistry, which Coleman du Pont visualized in 1902. Even as Pierre du Pont and his Executive Committee of 1914–18 directed the assembling and operation of the huge machine that fabricated 40 per cent of the powder fired by Allied guns, the company moved to conserve resources and man-power for the peace to come after the war.

New capital of $60,000,000 was authorized in 1915. It was primarily for the purpose of effecting expansion into new industrial lines wherein might be possible "the continued employment of many of our valuable men whose services might otherwise be lost with the termination of military demands." A substantial portion of earnings was withheld from distribution to stockholders for similar investments.

The Development Department investigated, not merely hundreds, but literally some thousands of businesses outside the explosives industry. This work, done under the direction of R. R. M. Carpenter, covered most of the industrial structure, parts of it in minute detail.

Followed a metamorphosis of Du Pont affairs swift, complete and far-reaching. After the signing of the Armistice, explosives became factually a subsidiary Du Pont interest. More promising lines included paints, varnishes, pigments, lacquers, coated textiles, acids and heavy chemicals, plastics, and a variety of fine chemicals. These were only the nucleus of the company that was being built. During the next two decades, by planned expansion, Du Pont's new interests were to spread like the branches of a growing tree, until they embraced more than 10,000 separate items of commerce, many of them previously unknown and non-existent.

Evidence of the effect upon the company is shown by its books. In 1913, to which status the business might easily have reverted, it employed 6,222 men and women, had $75,000,000 in assets, and derived about 97 per cent of its income from explosives. At the end of 1939, in contrast, Du Pont employed 54,800, owned assets in excess of $850,-000,000, and more than 90 per cent of its income came from sources other than explosives. Here is reflected a highly significant industrial development of our time.

As early as 1828, the German chemist, Woehler, synthesized urea, an organic substance, by chemically compounding inorganic substances. That sounds dryly unspectacular, but state it in less technical language and it becomes an exploit inspiring the imagination. What Woehler did was to create out of substances of non-living origin a material produced hitherto only in living bodies. The feat opened an infinite vista of possibilities. All that was needed to manufacture organic materials to order, or perhaps to create wholly new materials, were the scientific tools and the skill to use them efficiently and cheaply.

In England, in 1856, Perkin discovered that dyes could be made chemically from coal tar, which up to that time had been little more than a waste material. This was of tremendous importance because dyes were in common use. By 1872, Hyatt was manufacturing in a plant at Newark, New Jersey, the new ivory-like pyroxylin plastic he had named "Celluloid." In the late Eighties, Chardonnet introduced "artificial silk," forerunner of rayon; in 1909, the first plastic to be derived from coal was invented by Baekeland.

Within the span of a lifetime after Woehler's purely academic work with urea, factories in a dozen nations were manufacturing "man-made" materials that competed with such ancient natural materials as silk, ivory, amber, tortoise-shell, coral, horn, bone, the colors obtained from roots, bark, insects and plants, and even with rubber, leather and wood. The natural dyestuffs industry, almost as old as man himself, had practically ceased to exist in 1914; textile fibers of chemical origin were a commercial reality; the plastics industry was well established.

This scientific-industrial advance, speeded by such revolutionary forces as steam-power and electricity, had given birth to a new vision of the possibilities of industry in improving the living conditions of all people.

The belief that Nature is omnipotent, and that no material can be as good as that naturally produced, is one of the oldest of popular fallacies. Nature excels only in her own sphere. Her products, as man finds them, are seldom fully satisfactory as industrial raw materials. They are notoriously irregular in quality, heavy when the manufacturer would prefer them light, soft when they should be hard, solid when a liquid form would save unending expense. The chemically pure is rare in Nature's realm—substances are usually in combinations that are not easily broken up. The most useful materials are often in the least accessible places, exist in negligible quantities, or are the subject of national monopolies. Many badly needed materials have never been found in Nature at all.

The adapting of natural materials to modern uses, the expense involved in assembling them, the losses that result from their too early failures and inadequacies in service, to say nothing of the ravages of rust, rot, pests, pestilence, flood, drought, heat, cold and storms, add billions to the cost of all goods and services. Here is the biggest hole in man's pocket. Through it slips, as a conservative guess, not less than a dollar for every dollar he spends constructively on living. Patch that hole and the poorest family might double its possessions!

In 1914, Germany had gone farther than any other nation toward patching that hole, by applying science to her industry.

She alone possessed industrial scientific research machinery on a huge scale. Creation of new materials was its major aim. The result, when war stopped the flow of German-made goods to the rest of the world, was disclosure of an alarming dependence upon German chemical factories for certain essential materials, the equivalents of which Nature had ceased to supply.

Germany had been manufacturing more than three-fourths of the coal-tar dyes in use and supplying intermediates for most of the remainder. In England, where the secret of dye synthesis had been discovered, dyes were lacking even for uniforms and flags. The situation in the United States was so desperate that the Government urged Germany to sell us enough colors to print our stamps and paper money, and sought England's permission to bring the colors through the blockade in Dutch ships. The American textile, leather, paper, paint, and ink industries were disorganized by the dyes shortage.

Germany controlled the supply of many important medicinal drugs, anæsthetics and disinfectants upon which physicians and hospitals of the world had grown dependent. These, synthetics all, had been developed from coal tar and had no satisfactory substitutes. Moreover, Germany was the world's organic chemical center. Over a period of years, her universities had emphasized training in chemical synthesis; her laboratories had created thousands of new organic compounds, of which only a few relatively had been evaluated as yet. No other nation possessed these materials, from which, by all signs, there was in time to be built a new industrial world founded on science.

The United States had a thriving inorganic industry, but all the nation's coal-tar products plants together employed only 528 workers and were dependent on Germany for their principal supplies. America also relied upon Germany for potash. This, plus dependence upon Chile for nitrates, put the country in the dangerous position of being forced to go abroad for two of the three most essential fertilizer ingredients—phosphates alone were plentiful at home. The world's largest user of rubber, the United States, produced not one pound on its own soil. For silk, camphor, and many essential oils, American industry was dependent on the Orient.

This was only one phase of an unsatisfactory situation. Further expansion of all industry, and notably of such promising new industries as automobiles and airplanes, had become largely dependent upon the development of better, cheaper and more efficient materials. The motor car of 1914 was still little more than a costly plaything. The metals being used had been adapted from those designed for other purposes. Tires wore out after a few thousand miles. The car's painted finish cracked and dulled in a few months. The motor soon choked with gum and carbon. Rust ate into the machine's vitals.

Similar shortcomings were inherent in most things in wide daily use. The fault was not in lack of human skill, vision, or money, but in lack of the right kind of materials. Throughout industry this dearth was felt.

In the pinch of war, a popular realization developed of the importance of chemistry in modern life. The dyes shortage affected everybody. The absence of drugs made in Germany reached to the humblest sufferer. Poisongases and high explosives made plain that it was a chemical war and that a chemical industry was necessary to the security of every nation. The term "national self-sufficiency" began to be used. It was a vital goal yet to be attained.

Along every national boundary throughout the industrialized part of the globe tariff walls were thrown up, behind which chemical science might be nurtured. Great Britain joined officially with private enterprise in launching a British dyestuffs industry. Elsewhere in Europe direct subsidies, tax remissions and special privileges encouraged the erection of chemical plants as patriotic enterprises essential to each nation's life.

Close of the war intensified this race. Italy and Czecho-Slovakia deliberately promoted the formation of national chemical trusts that would be free of domestic competition. The chief chemical interests of Great Britain were merged into a single corporation, Imperal Chemical Industries, Limited. The old German Dye Cartel, which had been a voluntary pooling of patents and resources, was welded into the great Interessen Gemeinschaft Farbenindustrie Aktiengesellschaft—for short, the German I.G.—of semi-official status. Several years later, even Soviet Russia abandoned her antagonism to science and made creative scientific research a function of the state.

America's chemical future was left, as American tradition dictated it should be, to the initiative, vision and boldness of competitive private enterprise. Private enterprise rose to that challenge with the public at its back.

During the war years alone, more than $200,000,000 was invested in the dyestuffs industry by 118 American companies. Then and later, more than $300,000,000 was spent in augmenting the chemical educational facilities of American schools and colleges. The need for capital that could support industrial-research laboratories capable of competing with those of the foreign chemical trusts was met where necessary by mergers among groups of important non-competitive companies already in the field.

The post-war growth of the Du Pont organization, with its broadly diversified chemical interests, was directly in stride with this movement. Formed also were the almost as widely diversified Allied Chemical & Dye Corporation, and Union Carbide & Carbon Corporation; the American Cyanamid Company, Dow Chemical, Monsanto Chemical, and a long list of strongly financed, aggressive smaller companies. Research to improve, to reduce costs, and above all to create better and less costly materials and products, became more than the rule in this welter of post-war chemical competition in the United States. It became the means to business survival.

Scarcely an industry or a manufactured article has escaped the resultant revolution that has boiled from the test tubes of our chemists. Change has invaded our clothes-closets and kitchens, our medicine-cabinets and garages, our pocketbooks, and even our stomachs. The chemist has made rubber from coal, limestone and salt, the finest of velvets from wood, the rarest of perfumes from coal tar. He has captured the nitrogen of the air and the hydrogen of our streams to combine them with the carbon of our coal deposits to form new national wealth. He has excelled the silk-worm as a spinner of fibers.

Most important, he has created new industries and new employment, and provided the United States with a newly found security in the form of a dependable domestic supply of many essential materials, formerly obtained only from abroad. Lacking this supply, following the momentous events of December 1941, America would have been hopelessly lost.

"Venture Capital"

As we have seen, Du Pont set aside a large sum for non-military expansion between 1914 and 1918. This was only a portion of the amounts eventually to be invested "beyond explosives." Expenditures devoted to the discovery, development and large-scale manufacture of new chemical materials and products during the twenty years 1920–1939 exceeded $400,000,000.

More than 85 per cent of that total was spent after 1927. Better than $160,000,000 was disbursed during the depression-marked years of 1930–36. Despite the forlorn state of business in general during most of the decade of the 1930's, the Du Pont outlays on chemical research were on a steadily ascending scale.

Behind this effort, guiding it, and providing the financial support without which all progress in science and invention comes to a standstill, was a combination of men and purposes that it is pertinent to examine.

In his annual report for 1918, Pierre du Pont named sixteen men principally responsible for the company's wartime accomplishments. These were nine members of the Executive Committee, who had served through most of the war:

Irénée du Pont, Chairman, who was to all intents general
    manager
Harry Fletcher Brown, smokeless powder operations
R. R. M. Carpenter, development work
Frank L. Connable, special purchasing
William Coyne, sales
Lammot du Pont, miscellaneous manufacturing
Harry G. Haskell, explosives manufacturing
John J. Raskob, finance
Frank G. Tallman, purchases

Also seven heads of other chief departments:

Col. Edmund G. Buckner, military sales
Maj. William G. Ramsay (who died in 1916) and Harry M.
    Pierce, engineering
Dr. Charles L. Reese, research
Daniel Cauffiel, real estate
William A. Simonton, traffic
John P. Laffey, chief counsel

It will be noted that H. G. Haskell had replaced his
brother, Amory (whose death occurred in 1923). Hamilton Barksdale, another veteran of the pre-war expansion
period, had died only a few weeks before the Armistice.
That master of explosives salesmanship, Charles L. Patterson, had been in semi-retirement since the early days of
the war, but his nephew, Charles A. Patterson, was now
Director of the Explosives Manufacturing Department at
the age of forty-two.

With few exceptions, the leaders of the Du Pont wartime organization were men who, in 1914, were entering

the prime of life. They were then at the height of an aggressiveness that had been tempered by experience but not dulled by a plethora of it. A similar balance of seasoning with youthful spirit was to dominate the company's broadening activities of peace.

"I am firmly of the opinion," Pierre du Pont wrote the Board of Directors in April, 1919, "that we have now reached another turning-point in the conduct of the affairs of E. I. du Pont de Nemours & Company. Therefore, it would be wise to place the responsibility for further development and management of the business on the next line of men, advancing the members of the 'War Executive Committee' to administrative duties disconnected with the routine management of affairs."

He recommended that a new Executive Committee should be named:

Lammot du Pont, Chairman
F. Donaldson Brown
Frederick W. Pickard
A. Felix du Pont
Charles A. Patterson
William C. Spruance, Jr.
J. B. D. Edge
Walter S. Carpenter, Jr.
C. A. Meade

To prevent the loss of so much experience to the company, he proposed that the men who were retiring with the War Executive Committee should be elected to membership on the Finance Committee, if not already serving there.* His third suggestion was that Irénée du Pont should be advanced to the presidency.

Pierre, not yet fifty, was in the best of health. However, the directors, respecting his wishes, approved his program. On May 1st, 1919, he became chairman of the board, his brother became president, the new Executive Committee and the enlarged Finance Committee took office.

The youngest of the group, Walter S. Carpenter, Jr. (a brother of R. R. M.), was thirty-one. The oldest, Frank G. Tallman, was fifty-nine. Nor was all the youth concentrated in the junior committee. Lammot du Pont, thirty-nine, was a member of both committees. The Finance Committee also included Henry F. du Pont, thirty-nine; John J. Raskob, forty; the elder Carpenter, forty-two; and Irénée du Pont, forty-three. The emergency of the World War, which had catapulted young men into work of tremendous responsibility, was the reason for this remarkable dominance of "youth" in the executive superstructure of a company that was then 117 years old.

In the company's secondary executive structure was a similar dominance of younger men. Still in their thirties or early forties, they had participated in projects involving the employment of thousands of workers and expenditures of money in units of millions. The majority were technically trained. They saw the almost limitless possibilities of "science in industry."

*F. Donaldson Brown and Frederick W. Pickard had been elected to the "War Executive Committee" on October 30th, 1918, on which date also John J. Raskob resigned. Thus, with Lammot du Pont, three of the new committee had served with the retiring committee.

Especially important, these men were accustomed to working together as an organization. And more and more, in the years ahead, the emphasis in all American business activity was to be highly specialized, cooperative action. The individual was to be subordinated to the group, the group to the company, the company to the industry, the industry to the economic structure of which it was a dependent segment. The war had demonstrated not only that more could be done by organized effort, but also that the means for still greater organized effort were at hand in the form of better and more versatile facilities of transport, almost instant communication with every part of the continent, and power in abundance for large-scale factory operations. No such facilities had been available to any earlier generation.

On this stage, with these players, the Du Pont program, launched in 1902, continued to unfold, accelerated by three timely coincidences. First, the nation's vulnerable state caused by lack of an organic chemical industry stood both as a challenge and an opportunity. Second, most of the developments in which the United States was deficient —coal-tar products being a notable example—were related to explosives to the extent that the same or related raw materials were employed in their manufacture. The third coincidence was the presence in the company's councils of men so influential in shaping policies as John J. Raskob and R. R. M. Carpenter, and soon Walter S. Carpenter, Jr. None of them was a "powderman." Moreover, the three leading Du Ponts themselves—Pierre,

Irénée and Lammot—had interests afield from explosives. The country's economic security in peace, and in war, was one of them.

In 1915, the Arlington Company was acquired. It was an aggressive manufacturer of pyroxylin plastics, lacquers and enamels. The Fairfield Rubber Company, maker of rubber-coated fabrics, was purchased the following year to amplify the line of pyroxylin-coated fabrics added in 1910. One of the nation's oldest manufacturers of paints, varnishes and heavy chemicals—Harrison Brothers & Company—was bought in 1917, and at intervals these six other companies in the finishes industry were acquired:

Beckton Chemical Company
Cawley Clark & Company
Bridgeport Wood Finishing Company
Flint Varnish & Color Company
New England Oil, Paint & Varnish Company
Chicago Varnish Company

Upon the persistent recommendation of John J. Raskob, a 27.6 per cent interest was secured in General Motors Corporation.

"While there is no immediate relation between the explosives industry and the manufacture of motors," Pierre du Pont explained to stockholders, "this investment was made in such a way as to give opportunity for our financial organization to be of service, and at the same time increase greatly our financial strength."

Raskob urged the venture over strong opposition, at a time when investment in any automobile company involved risk. Only recently assembled, the units of General Motors had never been organized into a well-coordinated group. The financial structure was in need of buttressing. Pierre du Pont, Raskob, Donaldson Brown, and Alfred P. Sloan, Jr., who was then in charge of General Motors manufacturing operations, jointly undertook the task of laying the foundation of what, under Sloan's leadership later, was to become the highly efficient motors corporation of today. Research was given a place of first importance. So emphasized, it was to revolutionize the motors industry, especially in the low-price car field where comfort and appearance had been subordinated to utility.

Absorbed in this organizational work, Raskob and Donaldson Brown also were absorbed by it. They became vice-presidents active in the General Motors management, though they continued to serve with the Du Pont Board of Directors and Finance Committee. Otherwise, beginning in 1923, Du Pont participation in the motors corporation's affairs was limited chiefly to an advisory capacity on financial matters. The investment interest in General Motors, in 1941, was represented by ownership of about 23 per cent of the latter's issued common stock, then valued on the Du Pont books at approximately four times the $49,000,000 originally invested.

Amplified with new capital and expanded, the business bought outright opened jobs for Du Pont executive and technical talent, multiplying the opportunities open to ambitious men. In turn, the mounting opportunities attracted to Du Pont many of the best-qualified graduates of the country's technical schools, an invaluable addition.

Between 1925 and 1933, no less than nine important companies, each in a different field, were purchased. These, their chief interest, and the years of their acquisition were:

1925—The Viscoloid Company, pioneer producer of plastic combs, umbrella handles, toys and other plastic articles;

1926—National Ammonia Company;

1928—Grasselli Chemicals Company, since 1839 a leading manufacturer of acids and heavy chemicals;

1929—Krebs Pigment & Chemical Company, lithopone;

1929—Capes-Viscose, Inc., maker of cellulose bottle caps and bands;

1930—Roessler & Hasslacher Chemical Company, specialists in electro-chemicals, ceramic colors, sodium, peroxides, insecticides, etc.;

1931—Commercial Pigments Corporation, titanium pigments;

1931—The Newport Company, compounder of dyes and synthetic organic chemicals;

1933—Remington Arms Company.

The addition of Remington Arms was in the nature of a majority stock control only. The company was continued as a corporate entity under the presidency of Charles K. Davis, who had come to Du Pont in 1915 from the mining industry.

Thus far, the narrative has dealt only with the acquisition of existing enterprises. The latter were paid for, as a rule, in stock of the Du Pont Company, not in cash. The $400,000,000 program of development, already mentioned, was almost wholly *in addition* to these purchases.

Eight new manufacturing projects were begun in the following order:

1917—dyestuffs and other organic chemicals
1920—viscose rayon yarn
1923—"Cellophane" cellulose film
1924—synthetic ammonia
1924—photographic film
1925—industrial alcohol
1928—seed disinfectants
1928—acetate rayon yarn

The fact that a majority of Du Pont's more important original research developments date from 1928 and are not listed in the foregoing, is significant of the years of preliminary scientific effort they were to demand. The research expenditures over the years were to cost a great many millions of dollars; the expenditures for construction of new plants and additions to plants, to manufacture the products conceived or developed in the research laboratories, even more millions of dollars.

It is noteworthy that this money for construction was supplied, first, by the company chiefly from its undistributed earnings, or, second, by the investing public through the purchase of additional issues of Du Pont stock. In other words, it was "peacetime" money. Evidence of the benefits accruing from it is a statement in the company's annual report to stockholders for 1937. It said that new products developed largely after 1928—that is, during the depression years—represented 40 per cent of Du Pont's total sales volume in 1937, and gave work directly to 18,-

ooo men and women in employment that had not previously existed.

Four basic policies guided the expansion:

1. The industrial fields entered were still relatively new, or
else old but with prospects of improvement because of the very
lack of improvement marking their recent histories.

2. Du Pont-trained men invariably accompanied the investment of Du Pont dollars, by additions to the personnel of each
business acquired.

3. Usually, the investment was sufficient to insure ownership,
or at least control.

4. Investment of capital was consistently followed by the venture of more capital.

As the keystone principle of Du Pont management, that
fourth policy merits elaboration.

Investments are of two kinds, safe and venturous. The
former tend to maintain the *status quo* of industry. They
are based on "past performance" records and the belief
that the current routine of earnings will continue. Venturous investments, on the other hand, are founded on
faith in future accomplishments, on inventions yet to be
perfected, discoveries to be made, on growth and earnings only promised, perhaps mistakenly.

The Du Pont policy of expansion under which a "safe"
investment in a going enterprise was almost immediately
supplemented by a "venturous" investment aimed at discovery, improvement and new growth—hence new employment and new profits—was at once one of the oldest
principles of industrial progress and one of the newest.

It was new in the sense that, prior to 1915, few established manufacturers deliberately set out to find and develop new projects involving large "venture" investments with their consequent elements of risk. Instead, most conservative business men and bankers avoided risks unless the odds were long in their favor. Unproven projects were left to promoters, to the initiative and daring of individuals, and sometimes to the gullibility of the public; they cost the investing public huge sums when the promotion was unsound or dishonest. The manufacturer was rare who made it a policy to set aside a share of his earnings for the one purpose of experimentation, and later for the practical development of promising discoveries.

Those who insist that organized research under the auspices of large corporations is tending to eliminate the "lone inventor," usually overlook the fact that most of the projects of organized research are of a nature demanding expenditures far beyond the resources of any individual experimenter, and that only one experiment in ten results successfully. They overlook, too, that one of the greatest reasons for progress lagging in the past, as compared with its modern pace, was the inability of the "lone inventor" to get financial backing when he most needed it, during the doubtful stages of his work. They overlook the "starvation" period of invention, under the old system, and the frequent loss by the inventor of all rights in his discovery, simply because those rights had to be sold to obtain funds to continue his experiments, or to develop the result of them to the point where it was of practical value.

The Du Pont policy made possible scientific exploration on a scale unprecedented prior to 1914 outside of Germany. It assured the research worker of a sure, steady salary even though his experiments failed; and if his experiments succeeded, he received a commensurate reward. The policy provided the industrial scientist with the best equipment available, with able associates and assistants and careful technical direction, all serving to minimize delays and uncertainties and to reduce the chances of failure. Furthermore, the policy reduced the chances of loss by the investing public, because the Du Pont eggs were placed in many baskets, each of which was watched by a financially responsible company.

One other innovation should be emphasized. Late in 1921, the company's plan of management was changed drastically. Up to then, organization was along the lines of most American businesses. The Production Department did all manufacturing, the Sales Department did all selling, the Purchasing Department did all buying. Members of the Executive Committee, or "cabinet," personally managed these and other important departments.

The new plan set up five manufacturing departments and eight general or "staff" departments, each of which was demarcated to activities closely related. All manufacture and sales of plastics, for example, were concentrated under a General Manager. He was given full authority and responsibility for results in his own department. The departments that embraced general company services such as chemical research, development work, legal counsel, engineering, and so on, were placed in charge of Directors.

The change divorced members of the Executive Committee from the departments, except in an advisory capacity. They were, wrote President Irénée du Pont, "to be free to give all their time and effort to the business of the company as a whole," to the end that they "will be able to consider all questions and problems without bias or prejudice." The president was to be chairman of the committee.

Thus a dual vision was established in the upper management: one, the "whole company" viewpoint; the other, the specialized departmental viewpoint. The former served to coordinate the company's varied interests, and to keep its general policies in accord with the rapidly changing times. The latter saw in terms of specific industry's needs, and of specific practical problems in a particular field. In turn, departments were subdivided into still more highly specialized Divisions, over which were placed Division Managers.

With some changes and additions that have increased the number of manufacturing departments to ten and the number of general or staff departments to eleven, making twenty-one departments in all, this was the Du Pont plan of organization in 1942.

The new Executive Committee, as listed in the Annual Report for 1921, included Harry Fletcher Brown, William Coyne, Frank G. Tallman, and Irénée and Lammot du Pont, all members of Pierre's wartime committee. William

C. Spruance and Walter S. Carpenter, Jr., then Treasurer, were the others listed. Frederick W. Pickard was to return to the committee in 1924, after serving with the general manager group, and R. R. M. Carpenter was to return in 1925, after similar work, to relieve Tallman who retired. The calling back to the "front line" of so many of Pierre's veterans was necessitated by the fact that the reorganization, in effect, had created new posts in the upper management demanding the broadest kind of business experience and vision.

These men made up Irénée du Pont's cabinet for the balance of his presidency. The period was one of "growing-pains," technical and other difficulties, and problems of rising size and complexity, as we shall see. Meanwhile, in 1924, Dr. Charles L. Reese had retired as Chemical Director after a service with Du Pont of twenty-two years, though he continued active as a consultant until the end of 1930. His successor as Chemical Director was Dr. C. M. A. Stine, a New Englander who had come to the company in 1907 by way of Gettysburg College, the University of Chicago, and Johns Hopkins. The "third-line" men of war days were pushing to the front.

In 1926, Lammot du Pont became president of the company, Irénée du Pont vice-chairman of the Board of Directors. Youngest of the three brothers, Lammot was the best informed on the diverse new chemical manufacturing activities. Until 1929, he retained his brother's Executive Committee with but one change. William P. Allen was elected to membership from the general manager group, but, because of poor health, was able to serve less than a year.

Then a new Executive Committee began to form. The names of J. Thompson Brown, brother of Donaldson Brown, Jasper E. Crane and W. F. Harrington appear in the list given in Lammot du Pont's Annual Report for 1929. Two more new names, Angus B. Echols and Dr. C. M. A. Stine,† appear for 1930. They replace William Coyne and Harry Fletcher Brown, retired.

The next year, 1931, R. R. M. Carpenter retired. He was followed into retirement, in 1935, by Frederick W. Pickard, who had been a Du Pont vice-president and sales executive for fifteen years. John W. McCoy succeeded him, coming from the general manager group.

T. S. Grasselli, grandson of the founder of the Grasselli company, joined the committee in January, 1936, and served until the autumn of 1939. Shortly before that Henry Belin du Pont, aged forty-one, a grandson of the first Lammot du Pont and a great-great-grandson of the company's founder, was elected a vice-president and member of the Executive Committee.

Later we will take a closer look at this Executive Committee of the present day, and also at the Finance Committee. Sufficient for our purpose at the moment is to point out that while individual tenure of membership on the Executive Committee has been fairly long, as a whole the committee has received repeated infusions of new blood, and this new blood has been nurtured inside the company.

†Dr. E. K. Bolton succeeded Doctor Stine as Chemical Director.

The changes in the committee, moreover, have been frequent enough not only to maintain about an even balance between younger and older members, but in the course of two cycles of roughly fifteen years each, to reconstitute the company's active leadership in almost exact synchronization with the changing times in business. Meanwhile, the company's president, acting as chairman of the committee, has served as a continuing factor in the preservation of basic Du Pont policies more than a century old.

Suppose, now, we watch those policies at work on what, a brief span of years ago, was one of the most serious of all problems confronting the United States.

Dyes

THE dyes shortage, alarming to a dozen or more great American industries during 1914–18, was peculiarly of interest to the Du Ponts. They were already engaged in a most important phase of dye chemistry. Before dyes, or medicines, can be made from coal tar, it is necessary to produce from the black, sticky mass a small group of colorless liquids or white solids known as "crudes." Next, a secondary or intermediate group of chemicals, which may number hundreds, must be compounded from the crudes. Dyes are made from these "intermediates," which also enter into the production of innumerable products more or less remotely related to dyestuffs.

Interest of the Du Ponts in the dyes emergency centered in the fact that the explosives T.N.T., T.N.X., and picric acid, all stem from benzol and the other basic coal-tar crudes. Moreover, in the making of diphenylamine, used to stabilize smokeless powder, the chemists had to compound aniline from benzol—and aniline is a magic word to the dye chemist. From this colorless or brownish oil he derives the rich blue of indigo and a vast rainbow of other colors comprising one of the greatest of all dye families, the aniline dyes.

In view of this, the Du Pont Development Department began early in 1915 to explore the desirability of engaging in dye manufacture. At the same time, the few existing dye plants in America appealed to Du Pont for intermediates they could get no longer from abroad. From dye-starved textile, leather, and paper companies rose the demand for finished colors.

To many people, in 1915, it seemed only a step from explosives to dyestuffs. However, already the shortage of good organic chemists was acute; the company's technical staff was overworked. To enter any phase of dye-making meant venturing into one of the most intricately difficult of all chemical fields mostly with young, inexperienced technicians. That was only one of the problems that had to be faced by Du Pont management.

The prospect of profits from dyes was none too alluring. The annual consumption of dyes in the United States prior to 1914 had amounted to less than 25 cents per capita. This had been made up of hundreds of small items, the manufacture of which did not fit into existing mass-production methods. A huge investment was indicated according to the Development Department's studies, owing to the necessity of interchanging from one complex operation to another, raw materials, intermediates, semi-finished products and finished goods, and making every possible use of by-products. It was certain that the new industry could not result in anything but heavy losses for a period of years. It was almost as certain that, unless adequate tariff protection was provided, the country would be flooded with German-made dyes the instant the war ended and international trade was resumed.

Affirmatively, however, the United States needed a coal-tar chemical industry. Du Pont's logical trend of expansion was toward dyes. The Germans had made dyes pay profits. Washington was urging the development as essential to the nation's security.

"We ought to be as smart as the Germans," decided Irénée du Pont. "Let's see if we are."

In March, 1916, a group of Du Pont chemists began research on dye intermediates. As rapidly as they could be recruited, other chemists joined the unit. It became the Organic Division of the Chemical Department, and shortly numbered several hundred workers. On February 23rd, 1917, the Executive Committee appropriated $600,-000 for the immediate construction of a synthetic indigo plant on American soil, and the Organic Chemicals Division became an industrial department.

Lammot du Pont was transferred from black powder operations and placed in general charge of dyes, paints, plastics, and miscellaneous chemicals. These interests then weighed heaviest in the Du Pont planned future. Deepwater Point, opposite Wilmington on the New Jersey side of the Delaware River, was selected as the site for the dyes development.

The name was prophetic. A total investment of $7,000,-000 was contemplated. More than treble that sum was to be invested in dyes during the following five years. Not until eighteen years after inception of the venture—after

$43,000,000 had been risked—were the aggregate earnings from dyes and related chemicals to offset the accumulated losses incurred at Deepwater Point.

It was a unique adventure in new-industry building. Many of the recruits were fresh from the classroom. Textbooks had told them little or nothing of dye synthesis. Apparatus was lacking, and could not be bought. Factories for the making of heavy enamelware, special autoclaves, earthenware that would stand heat, and other highly specialized equipment, had yet to be built in the United States. Such equipment had to be improvised.

Nobody in the company then was experienced in the design, construction, or operation of dyes plants, which are mazes of technical complications. In England, the government had seized a German-owned indigo works, and to it a Du Pont mission gained access through an agreement with the English company that had taken over the works. Excepting for this help, however, ingenuity and knowledge acquired in other chemical fields were the only guides for the large-scale work under way.

Mistakes inevitably resulted. A plant erected for the manufacture of some urgently needed dye would often be made obsolete by the acquisition of new knowledge either by experience in operation or discovery in the laboratory. Sometimes a plant hurriedly erected to meet the insistent demands for a particular dye would not work at all. There was the constant conflict between the desire to progress cautiously and the contrary desire to supply quickly the dyes desperately needed by other industries. Heavy losses were the price of experience gained under these trying conditions.

All active German chemical patents on file at Washington were seized in 1918 under the Trading-with-the-Enemy Act. In 1919, they were sold to the Chemical Foundation, which had been organized to acquire and administer them. The Foundation granted licenses to all responsible American applicants on identical terms. However, out of a total of 4,802 patents, covering medicinals as well as dyes, only 403 were taken up by licensees over a four-year period.* Many of them were without commercial value or could not be made to work, because of lack of detailed disclosure, although the availability of the patents and security in their use did spur the development of many important colors.

At times, consequences were weird. Dyes would be synthesized in the laboratory, carried successfully through the stage of "semi-works" or small-scale experimental manufacture, and then placed in production. One day, the run of materials through the plant would produce a perfect product. Next day, using supposedly identical raw materials, formula, equipment, and operators, the resulting dye might be off in shade or wretchedly low on the yield obtained, for reasons then obscure.

The company's experimental expenses chargeable to dyes and dye intermediates stood at $2,364,000 in June, 1921. The country had sunk into a depression, which had

*Testimony of Francis P. Garvan, President, Chemical Foundation, Inc. before United States Senate Judiciary Sub-Committee, 1922.

reduced sales. Furthermore, American-made dyes had come into ill repute with consumers. Large quantities of poor colors had been dumped into a not-too-critical market during its years of urgency, when more than a hundred concerns had rushed into the business, many of them poorly prepared. Reaction had set in. The public was questioning the worth of American dyestuffs, good or bad. Dozens of the new dye companies had failed. In Du Pont's Executive Committee, more than one member was heartsick and beset by doubts.

Then, the dearly bought experience began, at last, to bring results. The first four or five of the fast vat dyes to be made in the United States were produced. This, in 1919, was encouraging. Successes with these and other dyes multiplied as the chemists mastered a system of testing and control more exacting than that demanded by any other chemical operation with which they had been familiar. The art of blending colors to achieve rigid and unvarying standards was mastered. Intensive, unrelaxing research on a steadily expanding front introduced one after another improvement in methods and products. By 1923, Deepwater Point was at the turning in the long dyestuffs lane.

Today, the Du Pont Company produces more than 500 separate dyes. Dyes of something over 700 different identities are now being synthesized in American factories. From these, unnumbered different shades can be produced by blending. In 1941, with the world again at war, the United States, excepting for a few novelties, was supplying all of its own civil and military requirements in dyes, in sharp contrast with 1914, when almost all dyes had to come from Europe.

So far as the quality of American dyes is concerned, no better are produced anywhere in the world.

Facilities for studying the proper use of the thousands of dyes that enter into commercial practice, and of testing them under conditions that approximate those of actual service, were among the earliest provided at the Deepwater Dye Works. The wide divergencies in the physical properties of dyed and printed materials, and the amazing diversity of uses to which these materials are put, constitute details that deluge a dye works technical staff. The ordinary street clothes of a man, for example, require several kinds of dyes irrespective of color, because cotton, wool, silk, leather, and felt are not usually colored with the same dye. Printing-inks make up a highly individualized division of color chemistry. Leather dyeing is equally specialized, as is paper dyeing.

Dyes are used to color almost everything. Intermediates employed in dye-making are the same or closely related to those used in the manufacture of pharmaceuticals, photographic chemicals, perfume and flavoring materials, rubber compounding materials, explosives, food preservatives, synthetic resins, and a host of other products. The Du Pont Company today makes more than 100 rubber chemicals, which taken alone as a group may be used to indicate the ubiquitous influence of synthetic organic chemicals. For example, in 1910 a good automobile tire would give about 2,500 miles of road service and cost about $25. Now a tire of the same size, but containing more rubber and capable of being driven at much higher speeds, usually renders between 15,000 and 30,000 miles of service, and costs initially only half as much as its 1910 predecessor. Organic chemicals made from dye intermediates are one chief reason.

Tetraethyl lead, use of which in gasoline was of primary importance in the development of high-compression motors, and which Du Pont manufactures for the Ethyl Gasoline Corporation, is another offshoot of the organic chemistry fostered in America by the dyestuffs industry.† So is synthetic camphor. Indispensable in the manufacture of pyroxylin plastics, natural camphor imported from Formosa and selling normally for about 50 cents a pound, reached the high price of $3.75 in 1918. The organic chemists at Deepwater Point replied by synthesizing camphor from the turpentine of Southern pine stumps, with the result that the price of industrial camphor sold in carload lots in 1939 was between 32 cents and 35 cents a pound.‡

The synthetic organic chemical industry, revolving around dyes, provides employment for a great body of sci-

†However, full credit for tetraethyl lead belongs to General Motors Research Laboratories. Du Pont's contribution to this development has been mainly in the field of manufacture.

‡Throughout the remainder of this narrative, 1939 figures are used preferably to those of later dates wherever they more accurately reflect "normal" business conditions in the United States, as distinguished from the "emergency" conditions that began to become evident late in 1940 and were pronounced in 1941.

entifically trained men, which, in turn, provides the incentive for the necessary academic training, an invaluable asset in the modern world. The number of doctorates in chemistry granted in the United States each year is now twenty or thirty times what it was in 1914, in large part because of this direct stimulation.

The Dye Works deals with a great variety of chemical reactions, which enable it to handle problems seemingly far afield from its own and cutting across the entire industrial structure. Its complicated network of relationships, built in a little over twenty years, touches almost all phases of American life.

IV

Cellulose Marches Again

WHILE Du Pont chemists stained their fingers with dyes at Deepwater Point, another group of experimenters at Parlin, New Jersey, ninety miles or so northward, struggled with a different problem in which color was a factor.

For centuries, men had been paying toll to the slow-drying qualities of paint. The toll had not been so serious in the leisurely days of hand-labor, but the rapid pace of mechanized industry had made it an item of formidable size on all cost sheets.

Motor cars and trucks rolled from the assembly lines of American automobile plants with the clanking monotony of a mechanical Mississippi, but at the toll-gate of paint, mass production groaned to a stop. It took at least a week to paint the humblest vehicle, a month or more to finish the motor aristocrats. No less than twenty-two separate coats of primers, surfacers, rough-stuff, glaze, color and varnish constituted the standard of painting perfection, such as it was, according to the methods of an art as old as the Pyramids and fundamentally as little changed. Acres of car bodies jammed warehouse floors in the process of finishing. Some producers had as many as 15,000 car bodies impounded at one time.

Furniture manufacturers were in a similar predicament, although they had learned to recognize that paint and patience were synonymous. Makers of pianos spent weeks in varnishing and polishing, coat upon coat. Locomotives and ships, bicycles and velocipedes, wheat-harvesting combines and tops, steel bridges and bric-à-brac, broomsticks and lead pencils, alike they waited upon paint and paid toll to its failings.

That toll, annually, ran into large sums. A glass of water carelessly placed on the piano marred in an hour the mirror-like surface that workmen had toiled over for weeks. The varnished tops of dining tables soon became ringed and blotched. Motor-car finishes cracked and discolored.

So acute was the problem in the automobile industry that special committees were named to make studies of it in the hope of finding a solution. Shabby-coated used cars, accumulating in dealers' stockrooms, jeopardized the sales structure and the industry's future. A body finish was needed that could be put on, dried, and polished in hours, and, equally important, which would withstand the punishments of road service as long as the car itself.

Such a finish, said paint experts, was as unattainable as the moon. It took time for paint to dry, and it always would, they asserted. The experimenters at Parlin, New Jersey, however, refused to be discouraged. They were not thinking in terms of paint. Lacquers occupied their thoughts.

Pyroxylin lacquers—roughly, a sort of "liquid cotton"—had become a part of the Du Pont line in 1905, with the acquisition of the International Smokeless Powder & Chemical Company, and the later addition in 1915 of the Arlington enamels and lacquers. They deposited a clear, glass-smooth film, which dried in a fraction of the time required by paint or varnish. Although excellent for keeping tarnish off polished brass doorknobs, beds, hinges, and other brass hardware, the lacquers of 1905 and 1915 were of negligible value for most other purposes. The possibility of compounding a fast-drying, durable lacquer for use on wood and other surfaces kept popping into the chemists' heads.

But failure followed failure. Heavier films were made, but they were too brittle. Nobody could find the secret of how to toughen them. The effort was no more spectacular than getting a new idea, testing it, disappointedly discarding it. Never say die is the chemist's slogan. So one hot summer day in 1920, a tough, hard film of lacquer was evolved that stuck to the handle of a hairbrush as if it had grown there. It had none of the faults of the "brass lacquers." It could be colored.

"Duco" pyroxylin lacquer, as the company trademarked the new finish, was to launch one of the swiftest rolling tides of change yet to burst from a laboratory test tube.

Three years were spent in testing and improving the product. Unheralded and unsung, it first went to market on brush handles and toys. Success was immediate, but toys were a long way from automobiles. Metal and wooden panels finished in "Duco" were set out on racks to be weather-tested through the four seasons. They were pounded, scratched and scrubbed, smeared with mud, sand, oil, grease, gasoline, alcohol and road tar. Used cars were lacquered and put into road service. Chemists sprayed their own cars and tested the finish at first hand.

All told, eight years were given to the research. At the end of that time, however, the Parlin experimenters were ready to move on to Detroit. They had facts, not surmises, to talk about. Their "Duco" lacquer dried in minutes. Heat would not crack the film. Water and mud would not injure it.

They went to a General Motors factory. One end of a floor was screened off. General Motors engineers worked side by side with the chemists through the stuffy heat of that summer. Difficulties attendant upon adapting the new finish to the exacting requirements of large-scale automobile production began to find solutions.

That autumn, the Oakland Division of General Motors introduced "Duco" lacquer on its open-car models and, as rapidly as production problems allowed, extended the use to the entire 1924 Oakland line. By April, 1924, Cadillac, Cleveland, Franklin, Lexington, Marmon, and Moon were prepared to furnish their cars with "Duco" upon request. Public demand for the new finish began to roll in upon retail car dealers, but still the bulk of the motor industry hesitated, was incredulous.

Then, in July of 1924, after exhaustive tests—one test-car had been sent around the world—Buick adopted "Duco" for its sports roadster and shortly after that placed it on other Buick models. The motor industry's paint dam burst. A majority of cars displayed at the 1925 shows specified the new finish as standard or optional.

Before another year, the old-style paints and varnishes were on the down-road to elimination from the field of automotive body finishes. Also the new lacquer was being used on washing-machines and gasoline pumps, on radio cabinets and golf clubs, on caskets and vacuum cleaners and furniture. Once under way, it swept the earth.

The paint industry was in a turmoil. Orthodox finishes could not compete with this sensational newcomer, which made it possible to finish a motor car in two days—later, in one day—and which wrote a new trade definition for durability and beauty.

The cost of refinishing new-car bodies damaged in shop-handling and in shipment had been enormous. A scratch or a rubbed spot had made it necessary to repaint the entire body section at a loss of a week or longer. With "Duco" lacquer, the marred surface could be rubbed down and refinished in trifling time. Dust-proof, heated drying-rooms had been required for the old slow-drying finishes. "Duco" made that factory space available for other uses. In humid weather, rejections because of faults in body finishes had been as high as 75 per cent. The new lacquer lowered rejections to 10 per cent or less.

The discovery did more than revolutionize methods; it changed the trend of thought. Progressive manufacturers everywhere sought to improve all types of finishes. The heart of the problem, they now saw, was creation of new finish materials. The original "Duco" lacquers were steadily improved. Fast-drying, easily applied enamels; brushing lacquers for use in the home, more durable and more colorful exterior paints were developed in rapid succession. Today, the finishes-compounder studies his industry's problems through scientifically critical eyes, which see paint as a first line of defense against decay, rust and general deterioration, and as a factor in the preservation of the national wealth.

Du Pont followed "Duco" with the "Dulux" series of synthetic resin enamels, introduced in 1927 and since then widely amplified.* The alkyd resins of "Dulux" finishes are formed by combining glycerin, oils, and phthalic anhydride, a derivative of coal tar, which means they are purely synthetic. Hundreds of forms of these resins may be compounded, ranging from viscous liquids to brittle solids. To the enamels made of them they impart unique properties.

A "Dulux" coating, for example, over metal or wood dries into a tough, flexible, air-tight, and also water-tight skin that even a sharp hammer-blow won't break. That skin, which is almost chip-proof, has the body and depth of baked porcelain. The drying speed of these enamels may be controlled, and they dry to a lustrous sheen that requires no polishing. They outwear old-style enamels by more than five times in tests with mechanical scrubbers.

In the refrigerator industry, and for kitchen cabinets, there was pressing need for a long-lasting, chip-proof finish

*The basic resin essential to "Dulux" enamels was originated by scientists of the General Electric Company, whose work Du Pont scientists amplified and carried to a commercial stage.

in white that would not turn yellow with age. The "Dulux" enamels provided a white that is whiter than white tile, and which stays white. Resistant to salt water and salt air, to corrosion and the abuses of the most punishing public services, these new enamels also are being put to multifold uses on exposed metal in heavy construction, in the shipping industry, and on buses, trucks, and streamlined trains.

Meanwhile, in the Du Pont laboratories and elsewhere, there is a wide, wide range of other materials under study which in the tomorrows to come may supplant the finishes now hailed as the ultimate. Change has become the order of the paint industry.

The Rayon Cinderella

O<small>N A</small> second front, cellulose marched during the decades of the 1920's and 1930's, and there, too, Du Pont interests expanded "beyond explosives." This front was rayon.

As the public saw it, rayon was without prestige in 1920, when Du Pont entered the business. Until its fortieth birthday, in 1924, it was without even a name. Cheap underwear and cheaper hosiery consumed most of the rayon output. Yet, in 1918, far-sighted investigators of the Du Pont Development Department reported, after careful study, that in the so-called "artificial silk" of the day was an infant industry that promised in time to attain major importance.

Negotiations for the purchase of American rights to European patents were conducted without result. Chemists were not available to launch research on a scale sufficient to establish the company in the industry through its own efforts, and it began to look as if expansion here might be seriously delayed. Then, unexpectedly, a large textile company's chief executive with French connections ventured a suggestion.

"It might pay you," he said, "to reopen this subject of artificial silk with the Comptoir de Textiles Artificiels, in France."

A Du Pont representative went to Paris. The Comptoir's attitude had undergone a surprising change. Subsequent conferences led to formation of the Du Pont Fibersilk Company, later to become the Du Pont Rayon Company, and still later the Rayon Department of the Du Pont Company, one of the largest of all Du Pont manufacturing departments.

A plant erected at Buffalo along French lines shortly was in production. This matter-of-fact business transaction, completed as the result of a friendly suggestion, was to have sensational consequences wholly apart from rayon. We shall come to them presently. Meanwhile, rayon was to prove a sensation in its own right.

Prior to 1911, Europe had the only successful rayon plants. Two attempts to manufacture in the United States had ended in bankruptcies. Then the English textile house of Courtauld, Ltd., built at Marcus Hook, Pennsylvania, the first plant of the American Viscose Corporation. Until 1920, this company was the lone producer in the American market. The Du Pont Company's erection of a competitive viscose process plant at Buffalo, New York, represented the first interest to be shown by a large chemical manufacturer in rayon. Moreover, as no other producer in the world had had so broad an experience with cellulose as a chemical raw material, or was so extensively engaged in cellulose research, the Du Pont invasion of the field helped to accelerate a readjustment of the industry's viewpoint that was to change rayon's character completely during the next decade.

The name rayon was adopted in 1924, and rayon chemists set out to blaze new trails. One of the first of them led to the ultimate consumer. Why not, the chemists asked, find out what the public wants and then make rayon to that order? They knew that, theoretically, rayon could be changed almost at will.

The consumer research machinery of the National Retail Drygoods Association was set into motion, American Viscose and Du Pont supplying the funds. Ten thousand women were interviewed. Later, Du Pont organized a division of textile specialists to conduct a continuing survey of consumer trends and to develop new fabrics in accord with them. The burden of the early reports was that low price was the chief impetus behind rayon's even then rising sales. But this impetus had its weakness. Rayon yarn prices were too high to permit the manufacture of low-cost goods without reducing the quantity of rayon yarn in the fabric. Consequently, even low-cost rayon goods were not giving people the kind of wear they had a right to expect.

Women objected to rayon's high luster, except for novelty effects, and asked for softer, heavier, runless, duller fabrics. American life had changed. The automobile had introduced women to the out-of-doors. Freed by new social concepts and new mechanical appliances from household drudgery, women had stepped out into a world of varied and vigorous activities.

This new world demanded new types of clothing materials: fabrics that would not soil readily, that could be easily cleaned, that would hold their shapes under all conditions, that were heavy enough for hard wear but also soft, flexible, comfortable in warm weather, smart in appearance, and reasonably priced. Wide gaps existed in the fabric range among cotton, wool, silk and linen. Modern living demanded that these gaps be bridged, that the range itself be broadened, that a textile assortment be created as varied as the activities of the time.

It was a huge order. The textile industry was already skidding into depression. Its morale and treasuries were depleted. Even if the needed new rayon yarns could be created, which was a theoretical possibility only, further costly experimental work would have to be done before new rayon fabrics could be developed. Neither fabric mill operators, dyers nor converters had money available for such a program. Rayon yarn producers faced the task alone.

No company, not even a nation, is due all credit for what followed in rayon. Every important rayon yarn-producing unit at home and abroad contributed something to the tidal wave of development that, rolling up during the latter half of the 1920's, broke over the entire textile field shortly after the turn of the decade.

Rayon was delustered. Its filaments, once ten times coarser than silk, were made finer than those of silk. The strength of rayon yarns was doubled. Yarns of twenty sizes, containing from eight to 180 filaments, were created. In all yarn sizes a uniformity of strength and quality, unknown in natural fibers, was attained.

By economies and improvements, rayon prices were

lowered from $2.05 a pound on 150-denier yarn in January, 1926, to 55 cents a pound in July, 1932. They have remained consistently within a few cents of that price since then, despite wide fluctuations in the price of natural fibers and other ascending costs. Wages paid by rayon yarn producers in the United States were advanced 65 per cent. Employment was doubled.

It was a resplendent Cinderella that emerged from the chemical kitchen of the rayon industry in the Thirties, one that was to dazzle and conquer and win astonishing acclaim. In 1937, the National Resources Committee classed the once lowly descendant of "Chardonnet Silk" along with the telephone, automobile, airplane, motion picture and radio as one of the six most outstanding technical achievements of the twentieth century. The Federal Trade Commission ruled that thenceforth all textile goods made wholly or partly of rayon should be so labeled, to end the growing popular confusion of rayon with silk, wool, cotton and linen!

There was sound reason for that edict of the Trade Committee. Rayon production in the United States in 1937 approximated 342,000,000 pounds. That was more than six times the consumption of silk, 97 per cent that of wool, and 9 per cent that of cotton. Nobody knew the exact figure, but the guess was probably conservative that not less than 100,000,000 pounds of this 1937 output reached consumers as some fiber other than rayon, because neither sales clerks nor customers could identify the true character of the synthetic fabric they handled. Moreover, this confusion existed more widely in the higher-priced fabrics than in the cheaper, and it occurred despite an intensive effort by Du Pont and other rayon manufacturers and retail stores to educate the public in the fact that rayon yarns no longer can be judged by the popular notions of a decade ago. Thus far had rayon come since 1920. A new textile Queen had been crowned!

The situation was fantastic. In the women's dress goods market, rayon was outselling silk by twelve pounds to one. In all feminine clothing, it was outselling wool, silk, and linen combined! The male wardrobe was being invaded. The magazine, *Men's Wear*, computed, as of August, 1937, that 76,500,000 pounds of rayon were entering annually into the manufacture of men's clothes and furnishings of all sorts, ranging from hat bands and neckwear to socks and garters; and added that rayon was being used in coat and vest linings to the almost complete exclusion of other textile fibers.

That year, too, a leading upholstery fabric for motor cars was a creation of rayon technicians. A new rayon cord of unusual tensile strength—a development of Du Pont laboratories—was doubling and trebling tire mileages on hard-duty heavy motor truck runs; in 1938, the same rayon cord was introduced in passenger-car tires. Rayon filaments that resemble horsehair and others that resemble straw were being produced for widely varied purposes, and since then the field of this man-made fiber of cellulose has continued to broaden. American rayon output in 1941 amounted to 573,230,000 pounds.

Being man-made, rayon is capable of continuing change and improvement. That fact is at once its forte and promise. And it also explains its rise, so fortunately foreseen by the Du Pont Development Department as early as 1918.

The American rayon yarn industry in 1941 was made up of sixteen producing companies, which operated twenty-odd plants in fourteen states, employed more than 50,000 workers, and represented an investment estimated to be in excess of $300,000,000. The Du Pont Company ranks second in yarn production in the American industry. Three Du Pont plants—at Buffalo, New York; Richmond, Virginia; and Old Hickory, Tennessee—are operated under the viscose process; one at Waynesboro, Virginia, established late in 1929, is devoted to the acetate process.

Now, let's look into the second development issuing from that friendly suggestion to reopen negotiations with the French Comptoir.

★ CHAPTER ★
VI

The Film That Sparkles

JACQUES EDWIN BRANDENBERGER, a Swiss chemist of Thaon, was a fastidious soul. He was shocked at the dirty tablecloths over which he saw hearty Gallic trenchermen eating in the Thaon cafés. Why, he asked, could not somebody make a cloth that would be impervious to the none-too-elegant table etiquette of the region, and from which bits of egg and splashes of good red wine might be wiped away by a whisk of the waiter's napkin?

The time was the early 1900's. As early as 1894, in England, Cross and Bevan had begun the manufacture of transparent cellulose sheets and film, having first covered their processes by patents. Later, another Englishman, Chorley, had devised additional processes of manufacturing cellulose sheets and film as an incident to his principal interest of the moment of transforming viscose into fiber.

Brandenberger decided to take up where the three Englishmen had left off; in 1908 he made a significant find. It was a somewhat brittle, sparkling, transparent cellulose sheet of no use whatsoever as a table covering, but in another field—the wrapping of gifts and other packages—he sensed possibilities.

He began to design machines to turn out the film in quantity and in continuous rolls, securing what patents he could in Europe and in the United States. Four years later, in 1912, he had perfected his machinery and had learned to make a thin flexible cellulose film for wrapping purposes, which proved surprisingly satisfactory. As the trademark, he coined the term "Cellophane," derived from the first syllable of cellulose, and the Greek word "phaneros," meaning clear.

The World War of 1914–18 intervened. Little thought was given to the commercialization of this new idea in packaging. Brandenberger's work went on, however, and by the time the guns were stilled he had nurtured "Cellophane" cellulose film into a stable product. Then he secured the important element without which no invention, however ingenious, can hope to succeed. He found a "backer" in the Comptoir de Textiles Artificiels.

Capital was placed at the disposal of the inventor. In 1920, a company called "La Cellophane" was organized. A showroom was opened on Rue de la Chausée d'Antin in Paris. "Cellophane" quickly won interest, but it was so expensive that it was demonstrated only in the most exotic uses, as on bottles of costly perfumes. By coincidence, it was during this period that the representative of the Du Pont Company went to Paris to discuss rayon with the Comptoir.

"We have," the Du Pont representative was told toward the close of the rayon negotiations, "a most promising new product of cellulose in which your company possibly may be interested. If our relations prove to be mutually desirable, we shall be glad to bring it officially to your company's attention as soon as our present transaction is concluded."

Relations did prove mutually satisfactory, with the consequence that, in 1923, Charles Gillet, representing the Comptoir, met with Colonel William C. Spruance, a Du Pont vice-president, in New York City. A Du Pont subsidiary company was formed to take over and develop the already established business of "La Cellophane" in North America, and to manufacture the cellulose film in this country, with title to the trademark "Cellophane."

Thus Brandenberger's sparkling discovery joined the growing parade of new Du Pont products. It shortly began to write industrial history.

The first cellulose film was known as "plain transparent." It was strong, protective, clear, and had what merchandising experts call "eye appeal." In 1924, with the inventor present, the first sheet to be produced commercially in the United States emerged from a new Du Pont plant in Buffalo, about the time "Duco" lacquer, its chemical cousin, was making its bow in automobile salesrooms.

The date was auspicious for an innovation in packaging. The "cracker barrel" era of merchandising had passed. Cardboard boxes, glass bottles, tin cans, tinfoil and other wrapping materials had come into wide use for all kinds of goods. The transparent film of beautiful name promised both protection and visibility for goods it covered, with versatility in addition. It was an innovation the retail trade wanted.

Although introduction of the new product to the American public was modest, the acceptance, once the true significance was grasped, was speedy. "Cellophane's" sparkle promoted sales. Despite the film's original selling price of $2.65 a pound, demand for it soon was strong enough to warrant enlarging the Buffalo plant, and broadening the merchandising plan.

As 1925 opened, the largest user of "Cellophane" was the high-grade boxed candy of S. F. Whitman & Sons of Philadelphia. That year, the price of the wrapping was reduced to $1.75 a pound—a slash of one-third—and cookies made by a Cleveland baker shortly blossomed out in cellulose film wrappers. This was the initial wedge in the baking field. The baker's sales increase was so notable that a larger cookie-maker in Detroit took up the new wrapping. Then a large national baking concern adopted it, first, to meet the local competition in Detroit, later on a national basis because "Cellophane" wrapping sold many more cookies.

In 1926, came another victory. Back of it was drama which began in peasant huts in the Bavarian Hills. The post-war German government forbade the slaughter of brood sows. In a little while Germany was overrun with little pigs, little pigs that went to market the world over. Reports came from Canada that the Loblaw chain stores there had been selling extraordinary quantities of bacon wrapped in individual packages of "Cellophane." Packers, worried about an enormous oversupply, were interested. Experimenting with units of bacon in cellulose film, they were astounded at the results. Americans took to "Cellophane"-wrapped bacon.

More consumption of "Cellophane" meant increased production and still lower costs. In 1927, its price was down to $1.45 a pound. As the cost receded, new fields opened. Textile goods and the less expensive candies came into the fold. The largest potential users, in foodstuffs, lagged, however. The reason was that the "plain transparent" film was so far behind other protective wrappings in moistureproofing that food wrapped in it dried out much too rapidly.

Du Pont chemists developed a process which moistureproofed the film, though that accomplishment was not the easily done trick it may appear. Immediately food which suffered from quick staleness joined the "Cellophane" parade. Perishable goods became interested. Cigars and cigarettes entered the now fast-growing procession, offering a "pocket humidor" that sealed in the tobacco's freshness and giving smokers a new appreciation of one of the South's greatest crops. "Cellophane" soon was standard equipment for most packaged cigarettes. Chewing-gum got sticky at the seashore, brittle in the mountains, in spite of all efforts to protect it, but moistureproof "Cellophane" pointed the way to a stable quality.

From the beginning each grade of cellulose film—plain transparent and moistureproof—has been sold at one price schedule, and the price has been lowered as rapidly as rising production schedules warranted. Twenty consecutive price reductions in a period of fifteen years dropped the cost of plain "Cellophane" to 12 per cent of its original cost; and between 1927 and 1939, the price of the moistureproof grade fell to 26 per cent of the original figure in fifteen always-descending steps. Meanwhile, quality and the ease of the film's mechanical application to packaging were improved just as consistently.

The original cellulose film plant at Buffalo, constructed under the eye of Brandenberger himself, has given rise to seven modern factories. Two are at Buffalo, New York; two at Richmond, Virginia; two at Old Hickory, Tennessee; and one at Clinton, Iowa. These factories supply "Cellophane" film to customers who do their own packaging, or to converters who process the film into bags, envelopes, printed sheets, ribbon and other forms.

Every day or so, some ingenious person stumbles upon a new use for the sparkling wrapping of apt name. The frozen-food industry, a prodigious growth arising from improved freezing processes and a demand for fresh produce the year around, has adopted "Cellophane" as an integral part of its product. The moistureproof wrapper has been found the most successful agent for protecting hard-frosted foods against dehydration and deterioration from air oxidation. The entrance of "Cellophane" into the kitchen has made an easier task of preserving home foods. The product's insulating properties, in combination with other characteristics, are commending it to the electrical industry. Decorators interested in a material offering unusual effects are making increasing use of it.

"Cellophane" cellulose film has become more than a new industry; it is an American institution, an integral part of modern life—a chemical commoner that "began at the top and worked down." This sparkling sanitary guard against contamination and waste is preventing losses in foodstuffs and other goods that might be reckoned by the millions of dollars yearly.

Plastics

ELLULOSE chemistry led Du Pont, in 1915, into plastics; and by 1940, plastics had led the company into the crisscrossing paths of an industrial wonderland maze.

At a banquet in Washington, in 1936, celebrating the founding of the American patent system, two apparently identical balls of pure crystal were displayed side by side on a lighted dais. Guards stood by, for one of the spheres—the eye could not distinguish which—was beyond valuation. The other ball was no less beautiful and as exquisitely wrought, but there all likeness to its rare companion ended. It was light and practically unbreakable. The material that composed it could be molded under heat and pressure into hundreds of useful articles, or it could be sawed, cut, turned, carved, drilled and polished. Carloads of it could be produced, as needed, from raw materials such as coal, air and water, either in crystal-clear form or in a rainbow of transparent, translucent and opaque colors, lastingly beautiful.

Rarity set the value of the genuine crystal ball. It was an ornament and no more. The other, fashioned from a Du Pont plastic at a cost of less than $50, symbolized an American industry that normally manufactures more than

$50,000,000 worth of goods yearly and gives employment to thousands. Here, in the tableau of the crystal spheres, one Nature's, one the chemist's, was presented the drama of Modern Plastics.

Plastics are chemical combinations of Nature's basic materials. Molded under pressure, they take and retain desired shapes, may be given almost any color effect. Numerous in character, almost incredibly versatile, plentiful, comparatively cheap, there is almost no end to the things that can be made from them.

The whiteness of alabaster, the translucent yellow of amber, the delicate veins of ivory and ebony's soft deep black, are transferred wholesale through plastics to fountain-pens and telephones, to bracelets and to housewives' brushes. Pearls and semi-precious stones are simulated in costume jewelry, and fashion has accepted the chemical legerdemain on its merits. At the same time, the plastics are filling needs that if dependent on natural sources to be met would constitute a serious lack. For example, there are no known natural materials that could be satisfactorily substituted for the several kinds of plastics employed in the manufacture of telephones, electrical devices of a wide range, and automobile laminated safety glass, to mention only outstanding instances.

A maker of grocery scales found that his salesmen had to be brawny as well as brainy to demonstrate their ware. Some metal counter scales weighed as much as 165 pounds. A plastic material was substituted for the metal. Not only was the weight of the 165-pound scales cut to 55 pounds, but the scales immediately were made available in new finishes and bright colors, as durable as the machine itself.

Gears that are as tough as steel but much more silent are being made of laminated plastic materials. The new Library of Congress contains $100,000 worth of these synthetics, used both for decoration and utility, and a comparable sum was spent on plastics for the de luxe cabins of the British liner *Queen Mary*. One must get down to minute details to measure the range of their usefulness in everyday life—the tips of one's shoe laces, for instance. The latter may look like metal or bear the finish of leather, but try a knife blade on those lace tips and you discover a plastic more often than not; and unlike metal tips, the plastic one's won't come loose. It has been estimated that in this use alone, in 1941, plastics saved 500,000 pounds of badly needed tin.

For decades Leominster, Massachusetts, was the chief horn-comb producing center of the world. Long-horned cattle that roamed the plains of the Americas, Africa, Asia and Australia in seemingly inexhaustible numbers were the source of its horn supply. From horn, Leominster's craftsmen also made hairpins and headdresses, some of them jeweled; buckles and buttons, knife handles and pipe stems, and the heads of canes and umbrellas. The meat of the longhorns was lean and tough, but on their wide-spreading horns they supported an industry.

The world scene shifted. Steam transportation ended the isolation of the plains. The refrigerator car was introduced. The food demands of growing populations made meat and wheat more important than horn, with the consequence that broad-rumped, hornless beef-cattle, sheep, hogs, and fields of waving grain pushed the longhorns ever back. As the century turned, the longhorns had all but vanished, and horn with them.

Horn products, once common, were becoming a luxury. So, too, were jobs in Leominster's horn factories when, in 1901, the Viscoloid Company was founded there to manufacture combs and other articles out of a pyroxylin plastic. Today, most of the world's combs are made of plastic materials, and Leominster continues to lead in their manufacture, as well as in the production of toothbrushes. Toys, jewelry, hair ornaments, and innumerable other things for which horn was poorly adapted are now made almost wholly of plastics. Plenty has sprung from dearth in the horn-products industry.

Stemming from cellulose on the one hand, and from coal or its by-products on the other, the two chief kinds of plastics are related in some degree to most of the products that Du Pont makes. Purchase in 1915 of the Arlington Company, makers of the pyroxylin plastic trademarked "Pyralin," was followed in 1925 by acquisition of the Viscoloid Company. From this nucleus was formed Du Pont's present Plastics Department.

The products of this department are intimately identified with the daily life of the average American. It turns out toothbrushes and combs by the tens of millions annually, as well as the bristles used in the brushes. It is one of the largest manufacturers of the plastic binder that has made possible laminated safety glass, and also of plastics from which are molded steering wheels, horn buttons, gear-shift knobs and kindred motor-car accessories. From the scuffless heels of her shoes to slide fasteners for her clothing and frames for her handbags and sun goggles, the Plastics Department helps dress the modern American woman. Legion is the number of articles made from the sheets, rods, tubes and molding powders sold to other fabricators in the plastics industry. Plastics have become essential in aviation at a time when aviation itself is vitally essential. They were indispensable parts of every bomber and fighting plane leaving American production lines in 1941.

Improvements in manufacturing methods have characteristically followed Du Pont into the business. Indeed, the entire industry has undergone a mechanical overhauling, inspired in no small degree by the quickened competition. For example, the plastic sheets used in making laminated safety glass were formerly sliced one by one from blocks in an operation as tedious as that of the old flat-bed printing-press. Now they are produced in continuous length by a remarkable machine developed for this special purpose. It is only one of many mechanical changes Du Pont engineers and chemists have fostered.

There are four groups of Du Pont plastics—the cellulose nitrate plastics, sold under the trademark "Pyralin"; the cellulose acetate plastics, designated as "Plastacele"; the "Lucite" methyl methacrylate resins, and the "Butacite" vinyl resins. All of these groups have certain properties in common, namely, ease of workability and adaptability to almost any form, shape, finish or color treatment. But each also possesses qualities that are unique, just as do oak and pine, aluminum and steel.

"Pyralin," in its heyday, is said to have had 25,000 uses. Wherever toughness and economy are primary considerations it is still the king of cellulose plastics. However, "Plastacele" is a better material where fire may be a hazard, as in lamp shades, millinery and airplane windshields. It is tasteless, odorless, and more resistant to the deteriorating effects of sunlight. In clear form, it is almost as clear as plate glass.

"Lucite" and "Butacite" are newer. Both are products of coal, water and air. The plastic crystal-like sphere of the Patent Office's dinner was made of "Lucite," which is clearer than optical glass and sets new standards for transparency and brilliancy in colors. You can read a newspaper through a block of it a foot thick. A solid rod of "Lucite" will carry light around bends or curves, a property which is leading to its use in dental and surgical instruments. A tongue depressor made of it, in which a light is placed at the base, lights the recesses of the mouth and throat. A similar instrument is used in examining the ear. The light is heatless because heat won't pass through the plastic material.

Discs are molded of "Lucite" that comprise a number of small prismatic surfaces which act on light like so many reflecting mirrors. When automobile headlights pick up these discs set in standards alongside highways, they become visible at surprisingly long distances. Employed along eighty miles of U. S. Route 16 between Lansing and Detroit, the disc reflectors contributed to a reduction of 79 per cent in night-driving accidents over a three-months trial period in 1938. The following year, they were on trial on the highways of twenty-four States. Today they are in general use.

Some of the vinyl resins that constitute the plastic trademarked "Butacite" by Du Pont, have been known to laboratory workers for years, but it was only by intensive research in which Du Pont participated with four other companies that very recently they were made available for industrial uses. The forms of the plastic may be varied from liquid adhesives to solids. Age, sunlight, heat and water have only a negligible effect on them. In certain solutions they combine readily with nitrocellulose, gums and other resins, which makes them valuable as finishes, especially on metal.

For years automobile makers were concerned over the failure of laminated safety glass to retain its full strength in below-freezing temperatures. Not only has "Butacite" removed that hazard but it gives new virtues to safety glass. The plastic is of such strength that its safety factor is unimpaired by extremes of weather. It is elastic. However, under the impact of a heavy blow, it does not spring but "eases" back to its original shape. Throughout the life of a motor car, it keeps its crystal-like clarity.

Plastics research, once directed toward achieving the one, ideal, all-purpose plastic material, now places its emphasis chiefly on developing specific plastics for specific needs, and counts these needs by the thousands.

VIII

Synthetic Rubber

ONCE more go back to the war of 1914–18. The major effort of Du Pont Company research during the two decades following was directed toward meeting deficiencies that became glaringly apparent during those years of conflict.

On July 9th, 1916, a gray submarine poked her nose out of the waters off Norfolk, Virginia, and headed for Baltimore to complete the initial leg of one of the boldest voyages of record. The daring visitor was the *Deutschland,* first underseas freighter. She brought badly needed dyes and drugs to the United States. She took back to her blockaded homeland nickel and 500 tons of rubber.

Lack of rubber had almost banished private automobiles from Germany's roads. Tires of wood covered with canvas or leather, and combinations of steel springs inside steel shoes, were being tried to meet the needs of heavy motor transport in the movement of troops, munitions, food and supplies. Worn-out tire casings were being patched and reused, filled with a chemical jelly instead of air, to save good rubber for gas masks, surgeon's gloves, airplane tires and storage batteries. Before the war ended, a rubber substitute of poor quality was selling for $30 a pound; rubber itself was practically priceless.

Germany's rubber shortage and its consequences awoke a world already dazzled by technical advances to the fact that rubber had become as indispensable as coal, iron, oil or copper. That fact was more alarmingly significant to the United States than to any other nation.

We were consuming, in 1918, approximately 160,000 tons of rubber annually. In twenty-odd years that consumption was to amount above 600,000 tons, or more than half of the world's rubber production. No other nation was so dependent on rubber, or had so far to go for it, or would be more seriously affected by interruption of its supply if it lacked a good rubber substitute. Yet the possibility of creating or finding any satisfactory substitute for rubber seemed remote when the Armistice ended the First World War.

Chemists had worked for nearly a century to analyze rubber's composition and to duplicate it by chemical synthesis. They had discovered that rubber's basic constituent is a substance called isoprene, and that a close chemical relative—methyl isoprene—seemed to afford the nearest practical approach to synthetic rubber. A reasonably good substitute for hard rubber could, in fact, be made from methyl isoprene, which, in turn, could be derived by fairly easy methods from acetone obtained from potatoes, grain or other starchy foodstuffs. But when the Germans, under war pressure, were forced to make automobile tires out of methyl rubber, the stuff proved a sorry substitute.

The English chemist, Tilden, declared that synthetic rubber was an impossibility "at any price." Rubber users in the United States pondered that fact with worried brows. In 1922, Great Britain passed the Stevenson Act to limit rubber production on her Far Eastern plantations. Prices began a wild ascent. In 1923, they passed 37 cents. In July, 1925, rubber was quoted in New York at $1.23 a pound, a prohibitive price.

The American motor industry was caught in the flush of expansion. Indeed, there was hardly an industry that did not feel the pinch of the restrictions. By unanimous vote, Congress appropriated $500,000 to survey possible rubber-producing areas in the Philippines and Latin America. Thomas A. Edison and experts of the Department of Agriculture started a nation-wide exploration of cactus, goldenrod, and other weeds and plants, for likely sources of rubber on American soil.

One historic happening went unnoticed. Quietly, in 1925, chemists of the Du Pont Company took up the search for a rubber chemically made. England's repeal of the Stevenson Act in November, 1928, found them still on the trail, one of the most trying upon which the company's experimenters had yet set forth.

Acetylene is composed of hydrogen and carbon, the two elements that also make up the isoprene unit of natural rubber. The Germans, during the war, had started off with acetylene to make rubber, but rubber had not been their primary objective. They had been making methyl isoprene, the base of their wartime synthetic rubber, chiefly from acetone derived from potatoes or grain by fermentation. Then, war conditions had rendered all foodstuffs scarce, and they had been forced to look for a new source of acetone. This they had found in acetylene, the raw materials of which—coal, limestone and water—were plentiful and cheap.

However, it took the German chemists four steps to synthesize acetone from acetylene, and two steps more to synthesize methyl isoprene. The Du Pont theory was that, by ignoring acetone, it would be possible to arrive *via* acetylene at another, as yet unknown, rubber base, in two steps only.

Pursuing this attack late in 1925, the chemists obtained diacetylene, and, by combining it with natural latex, were able to produce a rubbery material of a sort. However, late in December, synthetic rubber was as much a mirage as it had ever been. It was with little optimism that Dr. Elmer K. Bolton, Chemical Director of the company's Dyestuffs Department, which was conducting the research, left for Rochester, New York, to attend the First Organic Symposium of the American Chemical Society.*

Educated at Bucknell and Harvard universities, Doctor Bolton had completed his chemical training at the Kaiser Wilhelm Institut für Chemie in Berlin. There, in 1914 and 1915, he had gained first-hand evidence of how serious a shortage of rubber might become in a modern nation. He went to Rochester to hear a fellow explorer in acetylene, the Rev. Dr. Julius A. Nieuwland, C.S.C., professor of organic chemistry at the University of Notre Dame,

*Doctor Bolton was made Director of the company's Chemical Department in 1930, succeeding Dr. C. M. A. Stine, who then became a vice-president, a director and member of the Executive Committee.

read a paper dealing with the mysterious gas that had been his hobby since under-graduate days. Father Nieuwland was not even casually interested in synthetic rubber, but, in twenty-odd years of research, he had come to know more about the chemistry of acetylene than any other man in the world.

In scholarly manner, the chemist-priest reviewed his studies for the Rochester meeting. A layman might have been made drowsy by the ponderous chemical terms, but the Du Pont chemist listened with alert interest. Father Nieuwland had produced from acetylene, by using copper catalysts of his own discovery, a yellow oil, called di-vinyl-acetylene. During the reaction, he said, a strange odor had been noticeable, an odor that led him to suspect the presence, too, of mono-vinyl-acetylene, a gas.

The meeting broke up. The Du Pont man and the priest talked over the paper in more detail. Synthetic rubber was not mentioned. But seven months later, as a result of correspondence, another Du Pont chemist visited the priest at Notre Dame. An alliance was formed that allowed the Du Pont research group to continue the experiments with Father Nieuwland's catalysts.

More months of work followed, dangerous work, because acetylene is highly explosive. The yellow oil, di-vinyl-acetylene, was exhausted of every possibility as a rubber base. The months became years, until, in 1929, the elusive gas mono-vinyl-acetylene was produced in quantities that could readily be isolated and identified.

The chemists were growing "warmer" with each new

experiment. Shortly, they made their first great discovery. By adding to mono-vinyl-acetylene the gas hydrogen chloride, derived from salt, a thin, clear liquid was obtained. It was a wholly new material! The liquid was analogous structurally to isoprene, but possessed properties that made it differ decidedly to its advantage over isoprene. It was named chloroprene.

When properly treated, chloroprene polymerized into a rubbery substance several hundreds of times more rapidly than isoprene. Moreover, careful tests revealed that the resultant "chloroprene rubber"—they spoke of it as rubber—was not only as good as natural rubber for most purposes, but for many uses it was superior to natural rubber! In some it outlasted rubber!

The discovery pointed to the possibility of a commercial product that, once its manufacture in quantity was mastered, would make available to industry a new *kind* of rubber, one which would serve where natural rubber broke down. First offered under the trademark "Du Prene," but later given the generic name of neoprene, to designate it as an original discovery, this rubber was formally announced by the Du Pont Company in 1931, at a meeting of the American Chemical Society in Akron, Ohio. At that time a factory for the production of neoprene was already under construction at Deepwater Point.

It took courage to build a neoprene plant in 1931, apart from the intricate engineering problems such a project involved. The world's worst depression had set in. Natural rubber was selling at 5 cents a pound at New York.

Nine months were consumed in experimental manufacture. In the summer of 1932, at $1.05 a pound, neoprene went on sale in a market where values were being weighed more carefully than they had been in a generation.

However, the nation's processors of rubber gave attention. They found in neoprene a material that looks like crude rubber shipped from the rubber plantations, and that can be mixed with the same materials, processed by the same machinery, and vulcanized by the same methods as natural rubber.

Finished goods made of neoprene look like those made of rubber. They have the same elasticity, strength, and toughness. But the synthetic material is much superior to rubber in its ability to resist the action of oils, gasolines and chemicals. It is less affected by heat; is less permeable to gases. It is many times more resistant to checking and cracking when exposed to direct sunlight, and to ozone oxidation.

These characteristics made neoprene usable under conditions of service that placed rubber at a disadvantage despite its low price. In countless cases, the natural product was being employed most unsatisfactorily for the one reason that it was the only material of the kind available for the task. In still others, mechanical developments were actually being retarded because they demanded a rubber-like substance without rubber's faults, and it was non-existent until the advent of neoprene.

At $1.05 a pound, more than a half million pounds of

the new material were bought by the rubber-processing industry during the depression years of 1931–35 inclusive. Sales doubled each year, and kept on doubling. And with new plant facilities and steadily growing outputs, neoprene prices were lowered to $1.00, to 75 cents, to 65 cents, in keeping with the traditional Du Pont policy of sharing with its customers the benefits resulting from reduced costs.

By 1939, it was impossible to name an industry that was not making use of neoprene in many of its most important operations. Every automobile and airplane manufactured had neoprene among its parts. The only types of transport not using neoprene were the horse and buggy and the ox-cart. It was serving in every industrial plant. It was being used in place of casein, glue, gelatin, plastics, leather, pyroxylin, waxes, metals, fabric, and combinations of several of these materials. Moreover, this chemical offspring of coal, limestone, salt and water was giving excellent commercial service in solid tires for industrial trucks in the most punishing kinds of work, and experimentally was giving a creditable account of itself in pneumatic tires on highways and for military uses.

In the period 1934–1941, leading American tire manufacturers produced many tires made wholly of neoprene and tested them under a wide variety of conditions of service. Results were quite generally satisfactory. In addition, tires of all types and sizes were made with a neoprene tread and a rubber carcass. These tires also gave satisfactory accounts of themselves, and under some conditions of service proved to be outstandingly superior to tires made entirely of natural rubber.

When, in September, 1939, Germany's legions invaded Poland, neoprene output at Deepwater Point was at an annual rate of 3,000,000 pounds. New facilities nearing completion would bring production up to 6,000,000 pounds annually before the year's end. Viewing the blackening skies over Europe and listening to the rising agitation at home, Du Pont executives, in October, ordered still further expansion to give a total output of neoprene amounting to 13,000,000 pounds.

At the time, when one guess on what the future held was as good as any other but no better, this authorized expansion to more than four times the September capacity seemed a bold and possibly reckless step. It proved unexpectedly conservative. The sudden, swiftly rising demand from hundreds of American manufacturers that was precipitated by the country's defense program resulted in an acute neoprene shortage early in 1941.

The Government placed neoprene under mandatory priority on March 7th. The order was followed by Du Pont's announcement on May 7th that the company's Deepwater plant was once again being substantially enlarged and that a second and still larger plant would be constructed in Kentucky, with Du Pont funds, as rapidly as men and machines could build it.

The paradox of this situation is that the shortage of neoprene developed when there was, as yet, no similar acute deficiency in rubber supplies, the emergency that had been foreboded for years. When that emergency did become real, upon the outbreak of war with Japan, neoprene became doubly vital to the nation.

It would be foolhardy to say that this laboratory-born child of chemistry will serve equally as well as the natural product in all of the tens of thousands of uses to which natural rubber has been put. However, during the first ten years of its life, neoprene was adapted to thousands of purposes in which it demonstrated superiority over natural rubber in gruelling tests. That backlog of experience, as 1942 opened, was one of neoprene's greatest assets and promises of broader accomplishments yet to come.

★ CHAPTER ★
IX

Coal's New Scepter

A TECHNOLOGY known as high-pressure synthesis, which is as modern as this evening's newspaper, produces more than a hundred different chemical products from coal, water and air. They serve a thousand and one purposes—two of transcendent importance.

The specter of starvation threatening the civilized world because of a lack of nitrogenous fertilizers for crops, raised only a generation ago, has been laid for all time by the triumphs of high-pressure synthesis. We now can obtain all the nitrogen we need from the atmosphere, and its cost per ton is less than half that of nitrates derived from Chile's diminishing deposits in 1914.

The nation's military defense has been made secure in one of the most vital of all its essentials—nitrogen for explosives, to say nothing of steel, plastics and dyes, which also require nitrogen in their manufacture. The safety of the United States without a large assured source of nitric acid on home soil passed when the First World War established that a submarine can sink a battleship and that a single bombing-plane might close the Panama Canal, main Chilean gateway for the old nitrate fleet.

Today, we could lose the canal and still not suffer any serious inconveniences so far as nitrates are concerned.

That is a fact of no small magnitude. However, above it is another. There is good reason to ask if in gaining this one source of security, the chemists have not also discovered the source-springs of a vast new reservoir of wealth, so rich in possibilities that in time their find might be equivalent to the discovery of a new continent!

Alcohols for anti-freeze solutions such as are used commonly in motor cars; ammonia for refrigeration; light, crystal-clear plastics for aircraft, surgery, art, and innumerable other uses; solvents for use in lacquers; urea for fertilizers, medicines and plastics; these, together with the intermediates for nylon, and that first object of the development, nitric acid—are now only a few of the products of the high-pressure synthesis plant of the Du Pont Company at Belle, West Virginia.

Belle is situated not far from Charleston in the rich bituminous coal district of the Kanawha River. There, since 1925, an investment comparable to that of its great Deepwater Point Dye Works has been made by Du Pont in the construction of one of the most complex and fantastic of all chemical factories. Also, as in the case of the Dye Works, the Belle plant rose in answer to a foreign challenge.

Germany successfully "fixed" nitrogen from the air during the world conflict of 1914–18, after the naval battle of the Falkland Islands had decided that Great Britain, and not Germany, was to have access to the Chilean nitrate supply, and the Allied victory of the Marne had indicated a prolonged war. The chemical feat changed the ancient world picture on nitrates. Every major power sought to establish nitrogen fixation plants, as they became known. Muscle Shoals stands out as the great American governmental effort to this end, though the project was abandoned following the Armistice, in an incompleted state, and was not revived until comparatively recent years.

The early method of "fixing" nitrogen involved combining nitrogen and oxygen from the air in an electric arc at an extremely high temperature to form nitric oxides, from which nitric acid was produced. This process soon was to become obsolete because of the high power requirements. Better processes were developed, on the German pattern, to combine air nitrogen with hydrogen obtained from steam, or by liquefaction of coke-oven gas, to form ammonia. Nitrates were then made from the ammonia.

The Du Pont Company already had an excellent process for producing nitrates from ammonia, so independence and security lay in taking the first step and synthesizing ammonia from air and water. American rights to the Claude Process, developed in France, were acquired in 1924, and later rights to the Casale Process, developed in Italy. Construction at Belle got under way in 1925.

The facts that again a heavy investment was required, and that again there was an almost certain prospect of early losses, are readily understandable when the complexities of Belle are even casually studied. More than $27,000,000 was invested, by gradual additions over a period of ten years, before the cumulative yearly net operating results showed a dollar of profit. And that sum was greatly augmented by 1942, by which time the original process had been so improved as to be hardly recognizable in the modern operation.

In an endless stream day and night, 365 days in the year, coal—or carbon—is charged into one end of the Belle plant. From the other end emerges a long category of liquids and solids, offspring of the chemical union of coal, water and air.

The elements used are carbon, hydrogen, oxygen and nitrogen—one solid and three gases. These are the primary materials from which Nature creates vegetation, animals, people. They are also the raw materials of Belle. Synthesis at high pressures, therefore, is a man-contrived method of direct creation. Some call it the "chemistry of tomorrow."

Countless centuries elapsed before man learned how to use coal, his greatest heritage of stored energy. He knew that coal could keep him warm, turn his factory wheels and run his trains, ships and central power plants. However, he used only the energy value of coal until the Magical Age gave him a new vision. Now, he saw that coal meant not merely heat, light and power, but carbon, key element in no less than 500,000 known chemical substances. It is said that the possible total of carbon compounds runs into the billions. Thus, today's chemistry of coal is a technology of limitless frontiers.

Virtually the only raw materials entering the Du Pont high-pressure synthesis plant at Belle are coal, air and water. The daily coal consumption is sufficient to heat a large home for more than a century. The plant pumps enough water to supply a city the size of Washington, D. C. The gas works could furnish the requirements of Philadelphia.

The processing is so rapid that only thirty minutes elapse between gas generation from coke and the emergence of finished ammonia and crude alcohols. For the gases, however, those thirty minutes are a chemical "joy-ride" indeed. They are heated to 2,400 degrees Fahrenheit, washed and rewashed, plunged to 350 degrees below zero Fahrenheit. They are expanded to atmospheric pressure, compressed to 10,000 pounds to the square inch, and passed over various catalysts, those mysterious "chemical parsons" which aid in the unions of elements.

Considering the complexity of the processing and its extremes of heat, cold and pressures, it may not seem so strange that more than a hundred products are manufactured at Belle from three simple raw materials, and that these products cover practically the entire range of chemical and physical characteristics.

High pressure is, after all, entirely relative. In fire hose, water is propelled at a pressure of 125 pounds to the square inch. A modern super-pressure steam power plant generates steam at 1,300 pounds to the square inch. The Belle plant syntheses are conducted at square-inch pressures of about 9,000 to 15,000 pounds. "It is little wonder the gases finally combine," an observer once remarked, "if only out of sheer desperation."

Suppose, now, we look in more detail into this chemical labyrinth of Belle, which with all its devious turns is one of the most efficient operations of modern industry. Coal comes into the plant by carloads from mines at the plant's door. It is converted to coke by baking at high temperatures to drive off tar and gas. From the gas, solvents such as benzol and toluol are recovered—these are the "crudes" that also begin the synthesis of coal-tar dyes. What remains of the coke-oven gas is stripped of hydrogen and used as fuel; for nothing is wasted in this high-pressure technology.

The coke is pushed from the ovens red hot, quenched with water and delivered to the gas house. Here air is blown through ignited coke, the coke becomes white-hot, and gas rich in nitrogen and carbon dioxide is obtained. This is known as "blow gas." When steam is passed through the white-hot coke, the coke is cooled, and gas rich in hydrogen and carbon monoxide is produced. This is "blue gas." The alternate blowing of air and steam through the ignited coke—the one heating it, the other cooling it—produces the four gases from which ammonia, alcohols, urea, and the other products of high-pressure synthesis are derived.

Of the four principal gases produced in the gas plant, only hydrogen and nitrogen are required for ammonia. This clear, pungent chemical familiar to every housewife, is used as a fertilizer ingredient, for case-hardening steel, for huge-scale refrigerating units and ice manufacture, in oil refineries to prevent corrosion, in water purification, in cupro-ammonium rayon manufacture, wool

scouring, and in making glue, paper, pulp, pharmaceuticals, and ammonia salts. Then, of course, ammonia is rich
in nitrogen in a "fixed" or usable form readily transform-
able into nitric acid.

Urea comes next. It is formed when liquid ammonia
and liquid carbon dioxide combine under elevated temperature and pressure. Belle was the first operation to produce commercial urea in the United States. The product
is a white, crystalline substance, highly soluble in water
and alcohols. Historically, as we have seen, it is of interest
as the first compound of the animal kingdom to be synthesized from materials wholly of the mineral kingdom. It,
too, is widely used in fertilizers; the solid fertilizer compound contains 42 per cent of nitrogen. Urea's second
most important function is in producing plastics of the
urea-formaldehyde type, used for a vast miscellany of
purposes. Crease-proof and wrinkle-proof fabrics, a recent
advance of the textile industry, are based on urea. It is
further employed in the manufacture of adhesives, pharmaceuticals, animal foods, and in preparing finishes and
various types of stains.

Alcohols are made in equipment similar to that used
for ammonia. Different pressures, temperatures, catalyst,
and converter design are required, but the same "blue
gas" generated from steam and white-hot coke provides
the carbon monoxide and hydrogen needed for alcohol
synthesis. The first result, however, is not one alcohol but
a crude mixture of many alcohols and other organic compounds—a veritable chemical witch's brew. Distillation

achieves the desired pure compounds and fractions.

Methanol, or methyl alcohol, originally was produced in the United States entirely as a by-product of wood charcoal, hence the term "wood alcohol." Synthetic methanol was manufactured at Belle in 1927, in which year only about 1,000,000 gallons of the synthetic were produced in the United States. The output in 1941 was in excess of 32,000,000 gallons annually, and the price had been reduced 70 per cent. Uses are in the manufacture of formaldehyde, necessary constituent of two major classes of plastics; automobile anti-freeze, dyes, inks, wood and leather stains, cellulose nitrate plastics, and mirrors.

The alcohols may be converted to other organic chemical derivatives. For example, acetic acid and propionic acid are now in production. The acids combine with alcohols to form valuable solvents for use in varnishes, lacquers and other finishes. Too, on treatment with ammonia, they form amides, which are converted to amines. Certain amines are useful as flotation agents in the concentration of ores. You will encounter the diamines presently in the chemistry of nylon.

The lime salt of propionic acid is the active ingredient of an unusual product which suppresses mold and "rope" in bread and other food products. A few ounces of it in 100 pounds of flour is sufficient to maintain bread free from mold for a week or more.

Some alcohols, those of higher molecular weight, are used in mineral flotation, in the refining of beet sugar and in paper manufacture, as ingredients of automotive brake fluids, in plasticizers, as constituents of special lubricants, and as denaturants for tax-free commercial alcohol. Modern industry takes its many alcohols in copious drafts.

The story of high-pressure synthesis based on coal, water and air is a story of chemical invention, mechanical ingenuity, and of far-seeing business administration that affects the economy and security of the nation, the habits of millions of people, and the future of a giant industry. It is a story greater than that of Midas and gold, for, as Midas found, one cannot eat gold. With coal, water and air, it is different.

Already we wear clothing derived from those three chemical musketeers, and eat food from plants that subsist on their synthesized fertilizers. In several countries motorists are using synthetic gasoline made from coal. Not at all fantastic is the idea that some day people may live in houses made of coal-water-air plastics, and ride in automobiles made from the same three limitless materials.

X

More Tides of Change

HERE, let us summarize what has become a bewildering array of chemical progeny contributing to the welfare and security of the United States in peace, and also to its greater might in war.

Dyes and all the complicated miscellany of products stemming from coal tar constitute a vigorous American industry. The chemist has revolutionized the finishes industry, which also once hunted the earth over for many oils and resins now synthesized at home. An important replacement for rubber has been created from coal, limestone and salt, which the country possesses in abundance.

Only two nations—Japan and Germany, in that order— are larger producers of rayon, and the United States is dependent for rayon on neither of them. A thriving plastics industry is furnishing a galaxy of new materials, which are filling purposes once served only by the products of far-away lands. The technique of high-pressure synthesis is supplying not only nitrates for all needs, but an astonishing and growing list of other essential things as well.

In short, in two decades changes have been made in the world's trade routes vaster in number and scope than occurred in any whole century of earlier history, and in most of the changes chemistry has been the prime mover.

Yet, two great fields, heavy chemicals and electro-chemicals, have scarcely been mentioned. Not only have they
been decidedly important in the success of such notable
developments as dyes, rayon, cellulose film and neoprene,
but each has performed outstandingly in its own right.

The place of heavy chemicals—some of which, such as
sulphuric acid and soda ash, are manufactured on train-
load scales of production—was recognized early in the
Du Pont expansion program. Harrison Brothers & Company was bought in 1917, partly because of its facilities for the manufacture of sulphuric acid, which is
used directly or indirectly in virtually every industry.
These facilities were further expanded in 1928 by the
addition of the Grasselli Chemical Company, one of the
oldest and largest producers of heavy chemicals in the
United States. Later, the Grasselli Chemicals Department
was formed to take over the heavy chemicals end of the
Du Pont business.

This department is a leading modern representative of
the old inorganic chemical industry in the United States,
which even prior to 1914 was the world leader. Today, the
industry is still the leader in inorganic chemicals manufacture, but its advances have carried it also into the organic
chemicals field to an extent that makes obsolete its former
limited designation.

In the Du Pont structure, the Grasselli Chemicals Department is probably further removed than any other
from the public. The tons of sulphuric and other acids,
soda ash, caustic soda, phosphate salts, alum, barium

chloride, sodium silicate, and hundreds of other heavy
chemicals that roll from its sixteen strategically situated
plants usually enter industry through its back door. They
serve in pickling steel, making dyes, tanning leather, refining petroleum, in the complex metallurgy that in recent
years had produced 10,000 new alloys, to mention just a
few of thousands of applications.

Grasselli is one of the scene-builders on the backstage
of the industrial drama, which in one respect is unfortu-
nate, because heavy chemicals are an excellent show in
themselves. Here the ancient alchemist's Black Magic is
modernized to bring forth stuff as vital to progress as electric current. Much of it is "too dangerous to handle" by
other than experts, for it may be "volatile as gas, powerful as dynamite, poisonous as arsenic, and so odoriferous
that a bursting bubble in a witch's brew might be smelled
a block away," to quote a chemist. At the same time,
Grasselli produces things as mild as borax, as homely as
baking powder, and as salutary as Epsom salt.

During the Civil War and earlier, newspapers were
scarce, and circulation limited, but today, a thousand tons
of newsprint are used in the printing of a single Sunday
edition of a great New York newspaper, paper towels are
in public washrooms, many types of merchandise are sold
in paper containers. The United States not only produces
but uses ten million tons of paper each year, as much as
all the rest of the world combined. However, this is but a
fraction of the aggregate tonnage the paper industry con-
tributes to train, truck and water carriers. For every mil-

lion tons of finished goods that move out from the mills, about a million and a half tons of raw materials must be moved in, and these are largely heavy chemicals. They have made possible the modern paper industry. . . .

Wood preserving got its start with railroad ties. It is being extended to most uses of wood where permanency is desired in construction. The "heavy" research chemist has found that wood rot and stain are caused by fungi or parasitic plants, and he has developed economical chemical treatments for killing them. The result is that blue stain, once costing the lumber industry $10,000,000 a year, is a serious problem no longer, while the prediction is being made that rot-proof, termite-proof and practically fire-proof lumber will be used generally in building with the next generation. . . .

Allied research has also taken the front line in man's perennial warfare on pests, our most indefatigable and numerous common enemy. Pests try to destroy everything. They attack from the air, soil and water. Animals, plants, foods, clothing, buildings, furniture and ships are damaged or destroyed by them. Fortunately, a nature that destroys even as it creates keeps pests from taking complete possession of the earth, but the records of losses incurred yearly furnish alarming evidence that the natural processes are not effective either for man's full security or comfort. Damage done by rodents runs into the millions. Losses to shipping caused by marine growths on ship bottoms is put at 100 million dollars yearly. Insects alone are held responsible for an annual loss in the United

States of two billion dollars. The diseases and decay caused by fungi or molds and bacteria add a billion dollars more. Weeds take three billion dollars toll. Total these figures and you get an aggregate loss exceeding $6,000,000,000.

It was fairly simple in the earlier days to abandon pest-ridden areas, especially agricultural lands, and move on. Today, with most valuable land occupied, man has got to stand up and fight under conditions that more and more favor the pests. Federal and state governments, universities and private industry are spending large sums annually in studying the pests' life histories and habits and in searching for better weapons with which to combat them. Chemicals are a mainstay of the fight, including such heavy chemicals as sulphur, arsenate of lead, calcium arsenate, Paris green and other copper compounds, zinc chloride, the fluosilicates, and many, many others. Lately, through research in which Grasselli Pest Control Laboratories have been a leader, important new organic chemicals are being added to the list. . . .

Water-glass, to the average person, means a preservative for eggs. In a generation, however, employment of this heavy chemical, silicate of soda, has grown to include such uses as an ingredient in adhesives and soaps, the production of petroleum and its products; textile bleaching and processing; paper-making, wall-board, coatings for papers, wood and some metals; cement manufacture, and innumerable more minor uses. In twenty years the consumption of silicate of soda has more than doubled, and now amounts to about 600,000 tons annually. The soap

industry alone uses 150,000 tons per year, which is significant of a notable change in living.

In our grandfathers' time, soap was a negligible item
in the family's budget, and in industry's as well. Poor
grades of it could be had only at high cost. The average
family made its soap, largely from wood ashes and animal
fats. The "heavy" chemist has been an important factor in
making soap one of the cheapest and most widely employed of all manufactured products. In addition to silicate
of soda, he furnishes for it in quantity, at low cost, chemicals such as the alkali phosphates. Of these, "T.S.P.," or
trisodium phosphate, has largely replaced the more expensive sal soda which once was about the only major
industrial alkali to be had. Nearly 100,000 tons of "T.S.P."
are now consumed each year. . . .

Without sulphuric acid, the petroleum industry's phe-
nomenal growth would have been seriously retarded. . . .
Steel buys heavy chemicals in tank-car lots as science
enters more and ever more into the technology that is
creating lighter, stronger, more rust-resistant and longer-
lived metals. . . .

The use of alum and chlorine in conjunction with rapid
sand filtration of water has caused a reduction in the country's typhoid death rate from twenty persons per 100,000
in 1900 to less than one per 100,000 today, from which it
follows that protection is also present with that "chlorine
taste" in many drinking waters.

However, when one mentions chlorine, the province of

Grasselli is left behind, that of The R. & H. Chemicals Department is entered.

The Roessler & Hasslacher Chemical Company, founded in 1882, was acquired in 1930, and subsequently made The R. & H. Chemicals Department of the Du Pont Company. R. & H. furnished vital raw materials for the manufacture of dyes and tetraethyl lead, and also was, and still is, a leading factor in the field of electro-chemicals and specialized chemicals of a wide variety employed in electroplating, metal cleaning, bleaching, refrigeration, ceramics manufacture and innumerable purposes. The fact that the chief R. & H. operations are carried on at Niagara Falls, New York, is significant of the abundant electric power consumed in this department's principal processes, while the range of industries covered by its product is significant of the electrochemist's steadily expanding influence on American life.

A new method of chrome electroplating, developed by R. & H., normally can save the automobile industry alone $5,000,000 a year. Heretofore, the shell-like coating of metal on, for instance, the radiator of your car, was composed of a very thin layer of copper, a thicker one of nickel, and then the final coating of chrome. Copper was the basic coat, providing a bright smooth surface to which the nickel adhered, but the coating of copper had to be very thin, for it tended to become spongy when its thickness was increased. Thus the bulk of the shell was nickel.

The new process reverses this ratio. There is a thick layer of copper, now electroplated to provide a hard, tough finish, and a thin layer of nickel. The saving is effected by the fact that copper costs only about one-third as much as nickel, while the new process can be worked in about one-fourth the time required by its predecessor. Every article plated with chrome, from electric toasters to bicycles and banjos, is a beneficiary of this development. . . .

In 1919, the dry-cleaning of suits, dresses, overcoats, gloves was costly and consumed time. Naphtha and benzine were the chief cleaning fluids. They left a disagreeable odor. As they were both flammable and explosive, dry-cleaning plants were forced by city ordinances to locate in rural and suburban areas. The country's total business in dry-cleaning amounted only to $55,000,000 a year.

Today, Americans normally spend $600,000,000 a year to have their wardrobes dry-cleaned.* More than 200,000 local establishments and 12,000 plants serve an industry with a gross investment approaching a half-billion dollars. A freshly cleaned suit no longer advertises itself by its odor. It may be cleaned safely "while you wait." Costs have been sharply reduced. This sensational growth of the dry-cleaning industry was appreciably accelerated by new synthetic cleaning fluids, called chlorinated hydrocarbons, developed by R. & H. and other manufacturers.

These fluids also "degrease" metals, with such speed and efficiency that they are indispensable in the mass production of thousands of precision-built parts for machinery, ranging from typewriters to tanks. An indication

*Figures used in this paragraph are based on estimates made by the National Association of Cleaners and Dyers.

of their importance in industry is the fact that, by late 1941, the demand for them in essential military manufactures was so great that they virtually disappeared from the civilian dry-cleaning market. Thus all-encompassing are the demands of modern warfare. . . .

At Niagara Falls, the R. & H. Chemicals Department of Du Pont produces such diverse items as chloroform, weed killers, chemicals for the refrigeration industry, bleaches for textiles, powders for the cosmetic trade, disinfectants, insecticides, and a host of specialties essential in several score of industries. And simultaneously, at Perth Amboy, New Jersey, in a plant that is outside the main field of electro-chemicals, another R. & H. division works with gold, silver and platinum, and a gay array of colorful pigments and stains for the embellishment of china, glass, terra-cotta, tile, and all the varied products of a rejuvenated American ceramics industry.

Twenty years ago, fine ceramics manufacture belonged almost exclusively to Europe. The best dinnerware came from France, the reddest of pottery from China, and the beautiful pieces of English and German potters and glass-blowers were given favored places on shelves of American homes almost to the exclusion of home-made wares. That is true no longer. American craftsmanship has come into its own in the ceramic arts, and some of the most exquisite of the world's china and pottery and glass is now being made in the United States, with American materials and colors, and at much lower costs.

At one extreme, the new national independence is evident in the profusion of American-made ceramics in the displays of the most exclusive shops and the largest department stores. At the other, glassware banded in platinum, and china trimmed in real gold are being sold in "Fives-and Tens" and on basement counters. Moreover, because the life and amalgamated background of our western nation are so dissimilar to those of the Old World, the new American ceramic art is assuming a character of its own that is winning world admiration.

The early ceramic craftsmen knew little about chemistry. They mixed beautiful colors but had no idea why their mixtures gave certain shades; then found it difficult to duplicate a color tone exactly, or to develop matching pieces. The modern chemist has introduced science into the ceramic color art; he makes hundreds of combinations with one group of elements, varying the proportions in each case, to find which mixture gives the strongest and most pleasing color. He experiments with different degrees of heat, because some elements respond best to a low heat, others to a high one. Thus long strides have been made toward duplicating color results over and over with unvarying exactness.

In ancient days, too, color materials were limited to common elements. Craftsmen of those times had compounds of cobalt, copper, chromium, lead, arsenic, and also quartz, lime and soda. Many of the elements which are now used for ceramic colors were unknown to them. Compounds of vanadium, for instance, and the once rare element, uranium, and molybdenum, cerium, and neodymium, have been available for ceramic color work only during very recent years.

In the Du Pont ceramic color plant at Perth Amboy, at least 10,000 matching colors can be developed, and about 700 new combinations are made every year. Thousands of samples of standard colors are kept on hand in powder form for use in manufacturing ceramic materials for porcelain blocks, small glass jars, terra-cotta, brick, and bathroom fittings. A service laboratory is equipped to duplicate on a small scale every operation that will take place when the colors are finally applied in the factory. A color-detecting machine reveals fine gradations of shade not visible to the naked eye so that each color meets the specific needs for which it is ordered. Meanwhile, age-old implements of the ceramic craft have become museum curiosities and there is almost no wonder-piece of ancient potteries but that can be reproduced with scientific exactness beyond detection of the human eye.

Nylon

RESEARCH is of two kinds, "fundamental" and "applied." The first, also known as "pure" research, is essentially a search for knowledge regardless of its immediate practical value, from which possibly the next generation, or century, might profit. The second involves the application to everyday problems, where possible, of the knowledge thus acquired.

Early in 1928, fundamental research was started by the Du Pont Company. This was a venture into a field that had been left almost wholly to the universities and other institutions of higher learning, to endowed laboratories, and to scientific dreamers. However, over a century and longer, that field of effort had supplied the foundation of applied research in industry and medicine.

In charge of a small group of carefully chosen technicians, Du Pont placed Dr. Wallace H. Carothers, described by the noted chemist, Roger Adams, as one of the most brilliant students ever granted the doctor's degree by the University of Illinois.

Son of an Iowa schoolteacher, Carothers was thirty-two. For four years he had been teaching organic chemistry, first at Illinois and later at Harvard. Music-loving, moody, a merciless self-critic, he already enjoyed an enviable reputation in the higher reaches of science. It was agreed that he should select his own projects in "pure" research, that at his disposal would be the finest equipment available, abundant funds.

For some time, Carothers mentally had been exploring a bold possibility. By a unique method of attack, he believed he might be able to rival Nature in one of the most amazing and baffling of all her chemical feats, the creation of "giant" molecules. Invisible even under the most powerful microscope, these infinitesimals of organic matter play a mighty rôle in the grand scheme of creation. They are the building blocks employed in the evolution of all plant and animal life.

The growth of trees, grass, rubber, cotton—and men—is largely a process of polymerization, or joining of molecules into groups known as polymers, which are the unseen giants of this mysterious realm. Carothers proposed to synthesize the "giants" known as the linear condensation polymers, in which the molecules are linked end to end in chains of great length and strength. Silk, wool and leather are among the many common natural materials having polymers of this type.

The soundness of Carothers' plan of attack, pondered for months, was demonstrated on April 16th, 1930. On that day, his chief associate, Dr. Julian W. Hill, observed that the compound he had been treating in a molecular still had undergone profound changes. Instead of the weak, crystalline substance of short-chain polymers with which he had started, he now had a material that was tough, horny and elastic. The change was clear evidence that long-chain polymers had been synthesized. Carothers named them "superpolymers."

About two weeks later, Hill made a second highly important observation. In removing a sample of the molten superpolymer from the still—the stuff resembled a heavy, colorless molasses—he saw that it could be drawn out into a fiber which, astonishingly enough, was pliable and not brittle as its origin suggested it should be. Hill drew the fiber taut between his hands. It stretched to several times its original length, at which point it became "fixed" or oriented. The chains of molecules, thus drawn into line, were surprisingly strong. The fiber was opaque, lustrous, silk-like.

It was suggested that the discovery might be useful as a textile fiber. Upon being tested by Hill, however, it melted readily; hot water softened it; commonly used cleaning fluids dissolved it. Only casually concerned about any practical value of the fiber, Carothers put it aside and began a new series of experiments.

They had been working with a class of compounds known as polyesters. Now another, the polyamides, were prepared for treatment in the molecular still, and then mixtures of polyesters and polyamides. New fibers were formed, academically of great interest, but still of little or no practical value. However, Carothers had satisfied himself that what he believed could be done in synthesizing "giant" molecules, actually could be performed almost at will.

He turned to the hydrocarbons of acetylene. Searchers for a synthetic rubber, who were shortly to astonish the scientific world, long had been exploring this chemical maze. Carothers paralleled their work, employing his own theories, and in a few years he and his associates succeeded in compounding a whole new family of hydrocarbons of varying importance. From them, among other things, Du Pont chemists presently synthesized a colorless, viscous oil with an odor of tonquin musk, highly prized ingredient of fine perfumes. It could be produced at a fraction of the cost of true musk, which is obtainable only from the rare musk-ox of Asia at almost prohibitive prices.

Meanwhile, urged on by his associates, Carothers' own interest slowly had been rising to the possibility that had fascinated them in 1930, when Hill had drawn that first silk-like fiber that should have been brittle but was not. Rayon, being a chemical transformation of cellulose, which in itself is fibrous, is not a truly synthetic fiber. As yet, no chemist anywhere in the world had been able to create a truly synthetic fiber capable of meeting all the exacting requirements of the textile industry. That fact was a challenge not only to "applied" chemistry, but also, Carothers now realized, it was a challenge to his theories on the synthesis of "giant" molecules.

Accepting the challenge, he conceived a more practical way to build polymers of the linear type. The "pure" research was narrowed to polyamides. The molecular still, which had involved tedious delays, was abandoned for this fresh attack. On May 23rd, 1934, still another "superpolymer" was synthesized by still another unique process which the brilliant theorist had worked out mentally to minute completeness.

And this time, it was Carothers himself who demonstrated how readily the "superpolymer" could be spun into fibers. He drew the hot viscous substance into a hypodermic needle, and from the needle squirted a tiny stream of it into the air. As one watched, the stream cooled into a gossamer-fine filament, as fine as those of a spider's web. "Cold drawn" and subjected to tests, the lustrous filament equaled in strength and pliability any of the textile fibers. It "stood up" under heat, washing and dry-cleaning. Here was a definite turning toward the practical of a project on pure theory that had already consumed expenditures in excess of a million dollars.

"Here," said Carothers, walking into the office of the Du Pont chemical director in charge, "is your synthetic textile fiber."

For two centuries man had envisioned such a discovery.

Continuing experiments, following in swift succession, soon developed a sequence of more than 100 "superpolymers" from the polyamides. Their origin was such plentifully available raw materials as coal, water, air and vegetable oils. Each "superpolymer" was slightly different, and each was identified in the laboratory with its distinctive number. From these, early in 1935, the sample labeled "66 polymer" was selected after careful study as the most promising for immediate investigation from the strictly practical standpoint. The dividing line between fundamental and applied research was thus crossed. A new group of chemists took over.

Engineers, physicists, metallurgists, chemists who were expert in cost accounting and the problems of manufacture, joined in the work. An experimental or "pilot" plant was erected for the small-scale manufacture of polyamides into "superpolymers" and the strong, glistening filaments that could be spun from them. Scores of possible uses were explored, and it was found that the "superpolymers" could also be cast into tough leathery films, bristlelike filaments, and other forms.

On October 27th, 1938, almost eleven years after the beginning of "pure" research, and after nearly four years of "applied" research, Du Pont publicly announced the development of a "group of new synthetic superpolymers" from which, among numerous applications, textile fibers could be spun of a strength-elasticity factor surpassing that of cotton, linen, wool, silk or rayon.

This new group of synthetic materials was given the generic name nylon. These materials, said the announcement, were "protein-like" in character; that is to say, they resemble in chemical composition animal products such as silk, wool, leather, hair and bristles. Otherwise, it was said, nylon possessed qualities unknown in any other material. It was added that hosiery of the sheerest grade successfully had been knitted of nylon yarn!

The plodding, formative stage of Carothers' "pure" research was behind. Nylon captured the imagination. Newspapers hailed the discovery as "one of the most important in the century of chemistry." A fiber superior to silk, the writers pointed out, would insure the Nation's independence of one of the most ancient and complete of all natural monopolies. Women of America could wear fine stockings made wholly in America, of American-made materials processed by American workmen!

Du Pont exhibited nylon stockings at the 1939 World's Fairs of New York and San Francisco. By coincidence, it was the fiftieth anniversary of the first public exhibition of "Chardonnet Silk" at the Paris Exposition of 1889. Nylon hosiery created a sensation when, on May 15th, 1940, the first limited quantities of it were placed on sale in the Nation's stores. During the next twelve months approximately 64,000,000 pairs of all-nylon hose were bought by American women, the demand vastly exceeding the supply.

Few if any major inventions have been so immediately successful as nylon. The first facilities for the manufacture of nylon "salts" at Belle, West Virginia, and of "flake" and yarn at Seaford, Delaware, were obviously inadequate even before completion. And their enlargement was likewise incomplete when it became necessary to double the Seaford yarn capacity by erection of a second yarn plant at Martinsville, Virginia. By the end of 1941, Du Pont had yarn plants in operation or building capable of producing more than two million miles of nylon yarn daily, and the indicated direct employment attributable to nylon was 3,500. In addition, some 400 independent textile mills were using the yarn when, without it, they might have been seriously embarrassed as a result of the embargo then existing on silk.

As of the same date, only three years after the announcement in 1938 of nylon's creation, most of the toothbrushes made in the United States, half of the hairbrushes, and scores of varieties of industrial and household brushes were being bristled with nylon monofilaments. It was being used for tennis and badminton racquet strings, catheters, surgical sutures, fishing-leader material and snells, shoemaker's bristle, nylon-wound musical strings, and other products.

Experimentally, as 1941 closed, nylon was being employed as wire insulation, as a cement, in plastic-like formed objects such as self-lubricating bearings for machinery, and in a form resembling leather. A multiplicity of textile applications other than hosiery had been found for the yarn. They range from umbrellas and undergarments to shower curtains and—underscore this one—parachutes for aviators.

The outbreak of war with Japan, which definitely ended the importation of Japanese silk, made the advent of nylon one of the best-timed in the long history of invention. Yet war had been farthest from the thoughts of those bold explorers led by Carothers when they plunged into the chemical unknown back in 1928.

"Where next with nylon?" is a question that is debated with keen interest by Du Pont managers. The proteins— and nylon is "protein-like"—are present in all living cells, both animal and vegetable, which suggests a useful range for the Carothers discovery that might conceivably embrace thousands of articles that are basically protein in composition.

Even Carothers did not see the broadness of the nylon horizon—or did he? The practical value of his work was only partially determined when, on April 29th, 1937, the scientific world was shocked by news of his sudden death. The loss caused by his untimely end, at the age of forty-one, was probably as great in size as the contributions that in life he had made to man's knowledge.

During his nine years with Du Pont, more than fifty patents covering discoveries of major importance were issued in the name of this brilliant pioneer of science. Singly or in collaboration with associates, he contributed an equal number of papers to scientific literature. One year before his death, his achievements were recognized by his election to the National Academy of Sciences. He was the first organic chemist in the employ of industry to be accorded that honor.

★ CHAPTER ★
XII

Management

CHANGES of primary importance occurred in the upper management of E. I. du Pont de Nemours & Company as that business, begun in 1802 on the former farm of Jacob Broom, entered its 139th year of continuous activity on the American scene.

Pierre S. du Pont, now seventy years old, retired as Chairman of the Board of Directors on May 20th, 1940. The same day, Irénée du Pont, sixty-four, retired as Vice-Chairman; Lammot du Pont, nearing sixty, resigned as President and became Chairman of the Board.

Walter S. Carpenter, Jr., fifty-two years old and for more than three decades a Du Pont Company employee, was elected President. Angus B. Echols, fifty, succeeded to Carpenter's former post of Chairman of the Finance Committee.

For the first time since 1834–37, when Antoine Bidermann, a son-in-law of Eleuthère Irénée du Pont de Nemours, directed the firm following the founder's death, a man other "than of the name" was the administrative chief of the Du Pont Company.

The action, commented Lammot du Pont, was indicative of a "time-honored philosophy of management." Older men properly made way for younger men who had been long in the company's service. Pierre du Pont had emphasized the same philosophy upon his own resignation of the presidency in 1919. So had Irénée du Pont in 1926. The changes, however, symbolized more than a philosophy. The second cycle of Du Pont growth was completed; a new cycle began in Mr. Carpenter's promotion.

The business had closed its first century still a "family" enterprise, by Du Ponts forthrightly owned and managed. Until then, the savings of individual Du Ponts had supplied all capital needs; in person, Du Ponts had directed every important activity. The main, almost the only, interest of the company, was explosives. This cycle ended with the death of Eugene du Pont, Sr., in 1902.

Entered, then, the three cousins, Coleman, Alfred I., and Pierre. They saw "beyond explosives," boldly introduced blood other than Du Pont into the ownership and management; and in less than thirty-eight years assets expanded more than seventyfold. Preceding pages have outlined the diversity of chemical products and the wide scope of interests developed during this period. It follows that it was a well-rounded, essentially complete institution of which, in May, 1940, Walter S. Carpenter, Jr., became administrative head.

The dream of 1902 had been realized, to an extent that not even the far-sighted Coleman had foreseen. More than 70,000 investors, including 4,000 employees and every Du Pont executive of consequence, now owned the company's stock. Moreover, the upper executive staff numbered at least fifty men, and just below this managerial superstructure were some scores who filled positions indispensable to the continued progress of the business. Here were the cabinet officers, senators, congressmen, judges, and bureau chiefs of what, since 1902, had become a working industrial democracy. These men were "the Du Ponts" of 1940, in so far as the company exercised influence on the Nation's life.

Physically, the company's structure was made up of seventy-nine plants in twenty-five states. Many of the originals of these plants, later expanded and improved, had been acquired through outright purchase as a part of the process of building rapidly the type of diversified chemical organization which could hold its own in world competition and also maintain American chemical independence whatever the emergency. However, it was significant that no important purchase of an outside company had been made in seven years, whereas during the seven years preceding 1933 no less than eight major purchases had been effected.

In short, the organization of which Carpenter became chief had become "stabilized," at least to the extent that the purchase of other companies no longer was a necessary factor in its plans. The impetus for growth was being supplied by the company's own developments, based on its own research and promotional efforts. New industries resulting directly from this original research included neoprene, the synthetic "tree gold" which in some respects is superior to rubber; and the sensational nylon development. The third cycle of Du Pont progress was already under way; retirement of the elder Du Ponts and the advent of a new president merely date-marked that fact.

Walter Carpenter was intimately familiar with what the company had been, what it had become, and the course it had charted for the future. Ten of his thirty-odd years with Du Pont had been spent in the Development Department; for twenty years he had served on the Executive Committee, almost as long on the Finance Committee. Perhaps it is redundant to add that, upon taking over Lammot du Pont's vacated post, Mr. Carpenter was administering policies he had helped, on the one hand, to preserve, and, on the other hand, to formulate. The day following his election, Du Pont rank and file settled back to work as calmly as if no change had been made. In fact, none had been made, either in the company as such or in its larger aims.

Lammot du Pont's Executive Committee—the "cabinet" of the president—continued in office with only Lammot absent. The names of these committeemen were given earlier, but it will be instructive to review them:

Walter S. Carpenter, Jr., Chairman
J. Thompson Brown
Jasper E. Crane
Henry B. du Pont
Angus B. Echols
W. F. Harrington
John W. McCoy
C. M. A. Stine

The careers of most of these men have been so strikingly

similar in general outline that a composite of the group might be said to typify the Du Pont high executive. Six were born and reared in small towns or minor cities—only Brown and Crane were born in large cities; Harrington and McCoy are natives of Delaware. All attended college, and all but one specialized in chemistry or a branch of engineering.

The composite or typical high executive has spent at least thirty years in the service of Du Pont, or of Du Pont and a company acquired by purchase. He was promoted to his first major post at about the age of thirty-eight, advanced rapidly thereafter. He is presently in his fifties, is married, has children, and resides in or near Wilmington, comfortably but with little ostentation. Travel, study, and first-hand contact have made him well acquainted with the United States, its industries, its problems, its people, and also with other parts of the world.

Go through the list of Du Pont general managers of manufacturing departments, directors of general departments, and managers of larger divisions and one finds that, with age allowances, they, too, fit generally into that composite. One finds also that they are unusually well qualified to discharge their respective duties, which should not be surprising in an industry so keenly competitive as the chemical industry.

The Finance Committee exercises Du Pont's highest executive functions. In control of the purse strings, it may veto any major project by the simple expedient of withholding funds. This committee, as the company's third cycle began in 1940, had five Du Ponts among its membership of ten, which was as follows:

Angus B. Echols, Chairman
Donaldson Brown
Walter S. Carpenter, Jr.
A. Felix du Pont
Henry F. du Pont
Irénée du Pont
Lammot du Pont
Pierre S. du Pont
Harry G. Haskell
John J. Raskob

The fact that the weight of majority opinion alone, wholly divorced from that of stockholdings, determines the Finance Committee's decisions is one reason why the business world concedes that Du Pont's financial management is sound and far-sighted. Equally important, undoubtedly, is each member's familiarity with the company and the broad potentialities of the chemical industry.

Of course, over both the Executive Committee and the Finance Committee is the Board of Directors, which, in May, 1940, comprised thirty-six members. The directors are the elected representatives of more than 60,000 common stockholders and, as such, hold the final vetoing power. However, every member of the Executive Committee and the Finance Committee is also a member of the Board of Directors; and so, too, on the date mentioned, were the heads of eight departments, namely: C. R. Mudge, Legal; William Richter, Fabrics & Finishes; Edmund G. Robinson, Organic Chemicals; Fin Sparre, Development; F. A. Wardenburg, Ammonia; E. B. Yancey,

Explosives; L. A. Yerkes, Rayon, in which department is included nylon and "Cellophane" film; and James B. Eliason, Treasurer.*

Thus, an overwhelming majority of the company's Board of Directors were actively engaged in the working management in 1940. Additionally, eight Directors were formerly identified with the working management, namely: William P. Allen, Donaldson Brown, H. Fletcher Brown, R. R. M. Carpenter, Charles Copeland, T. S. Grasselli, Frederick W. Pickard and Harry M. Pierce. Among the five Directors remaining, Francis B. Davis, Jr., Chairman and President of the United States Rubber Company, was also a former Du Pont official; Alfred P. Sloan, Jr., spoke for General Motors; Eugene du Pont, Eugene E. du Pont and William du Pont, Jr., were very much interested stockholders residing in Delaware.†

The company's Treasurer, James B. Eliason, was another son of Delaware who found a life's work in his State's largest business unit; and in May, 1940, he was elected also a member of the Board and a Vice-President. The

*C. R. Mudge retired as head of the Legal Department on September 11th, 1940, but continued to serve on the board. Harold C. Haskell, native of Maine and graduate of Tufts College and Harvard Law School, succeeded him as Director of the Legal Department. On June 16th, 1941, following the death of William P. Allen, Mr. Haskell was elected to the Board of Directors. He is not related to the Haskells mentioned earlier in the narrative.

†Charles Copeland and F. B. Davis, Jr., resigned as members of the Board of Directors on January 19, 1942. Lammot du Pont Copeland and Crawford Greenewalt were elected directors to fill the vacancies. In addition to their diversified experience in the company's activities, the two new members of the Board of Directors are substantial stockholders and through their connections have knowledge of the views of many other substantial stockholders.

Secretary, William F. Raskob, like his brother, John J. Raskob, joined Du Pont in 1902 as a stenographer.

Few businesses in the upper reaches of the Nation's register of private enterprises had more widely delegated executive authority or were freer of "absentee" control than was the Du Pont Company in this, the 139th year of its history. The old absolutism in ownership and management had disappeared.

The contrary was true of the company's policies, however. The basic principles of Eleuthère Irénée du Pont de Nemours still governed. Changing times and broader vision had modified them only to the degree necessary to fit new conditions and an expanded organization. Just as the founder recognized the justice of Workmen's Compensation almost a century before legislative enactments under that name, his descendants had established a pension plan in 1904 many years in advance of most industry. Supplementary plans were, in 1940, commanding a substantial share of the company's annual income, and had been over a period of years. The trend in this respect was one of continuing liberalization.

In 1939, a normally representative year, the parent corporation expended approximately $8,400,000 for bonuses to executives and other employees, and $8,600,000 for its employees' protection against financial loss from sickness and accident, for pensions, life insurance, vacations, measures for industrial health and safety, and the like. Additionally, almost $4,000,000 was paid out on similar accounts under various State and Federal laws. The total outlay of $21,000,000 for the benefit of employees was equivalent to about 23 per cent of the combined wage and salary rolls.

Today, if a Du Pont employee of one year or more continuous service is taken ill or injured, his regular earnings are continued for a period of at least three months, without his incurring any debt to the company. Du Pont shares the cost of accident and health insurance, pays all charges incident to Group Life Insurance. Annual vacations of two weeks, at full pay, are granted to wage-roll and salaried employees alike. Pensions, provision for which is made wholly by the company, are granted "to employees who become unable to work because of physical or mental incapacity after fifteen years or more of continuous service," or upon compulsory retirement at the age of sixty-five. The Du Pont Pension Fund, in 1940, exceeded $26,800,-000; more than 1,000 former employees were on the pension roll.

That year the company's Medical Division maintained a staff of more than 100 physicians, technicians, nurses and assistants. Their work comprises medical examinations upon employment and periodically thereafter, the administration of first aid in emergencies, and advisory consultations with any employee who applies. The cost is paid by the company. If examination discloses medical attention is required, the employee is advised to consult his physician. The Haskell Laboratory of Industrial Toxicology was established, in 1935, to develop methods of protecting employees in the manufacture, and customers in the use, of such of the company's products as may be toxic.

Ever since the first Irénée du Pont astonished workmen with the unusual design of his original powder-mills, the company has emphasized "safety" in all its operations. Measures to achieve it have become increasingly effective. The frequency rate of major injuries in the company's operations as a whole, in 1939, was 1.4, which compared with a rate of 7.4 for the chemical industry as a whole, and about 11.6 for all industry. The frequency rate represents the number of injuries resulting in loss of time per million hours worked.

Publicly, the Du Ponts say little on the subject of industrial relations, yet even a casual study of their business history reveals a remarkable absence of strikes and other labor troubles, and also a notable success in attracting and holding able men. That record is not accidental. It stems from traditions which the Du Ponts of 1940 were seeking earnestly to maintain despite growing complications in the national industrial life.

What might be called the philosophy of the company's management, as it is evidenced concretely in employee relationships, merits more analysis. At the outset of their stewardship, Coleman, Alfred I. and Pierre du Pont set about making room at the company's top, not for assistants, but for associates. They provided, moreover, that these associates should receive rewards more substantial than titles. The bonus system that they inaugurated became a continuing means of encouraging and rewarding exceptional effort. Distributions of Du Pont stock, following the principles adopted in 1905, have aggregated many millions of dollars. The company, too, has assisted important executives in the purchase of Du Pont stock. The result cannot be measured, but few if any modern corporations are better endowed than Du Pont with that intangible asset, high morale.

In practice, the value of the bonus plan is greater to the recipient than the monetary value alone. Awards are made by the Executive Committee. They are based on personal observation and recommendations by department and division managers. Awards are not made casually. Thus, the entire organization is canvassed regularly for productive talent. Managers, and the company's president as well, learn much about promising men; those chosen for awards know that their efforts have been authoritatively noted.

The tendency among Du Pont men to look upon their employment as a life-work, and the performance records made by the organization in emergencies such as the First World War, may be traced in no small part to the company's readiness to let ability establish its own ceiling, and to the real effort constantly under way to evaluate ability in terms of demonstrated accomplishment. The Du Ponts, both personally and as executives, have been singularly successful in winning and keeping respect, simply by being themselves under all conditions and on all occasions, and by asking no more of others than they, too, are willing to do.

Lastly, the company's leadership in a highly competitive, highly diversified industry, wherein the new of today may be out of date tomorrow and the unknown is constantly a challenge; wherein venture is the first rule of progress and frontiers are ever moving on ahead; that, too, draws and holds men and women to Du Pont. It accounts for the company's refusal to grow old, and for the Du Pont man's pride in being known as a Du Pont man.

The old explosives firm was ceaselessly confronted with competition, but none of it was so keen as that which is the routine experience of every diversified chemical manufacturer in America today. "The modern idea," says Lammot du Pont, "is that competition should be not only recognized, but welcomed, if it is fair." However, even granted strictly fair competitors, the chemical producer now operates under infinitely greater economic pressures than are found in most other industries and, what is more, accepts them as normal and desirable. Complacency does not thrive in such an atmosphere.

The competition comes from three fronts. First, is that of the known, often age-old articles which the new chemical progeny seek to excel; second, is the competition of similar chemical products; third, is rival research. Without warning, rival research may introduce a wholly new, vastly superior development, and it might come from any one of a thousand laboratories anywhere in the world! This means that ceaseless research is imperative to the chemical manufacturer. Moreover, the quality of his research must be as high as, or higher than, the research of his competitors. It means diversification, both of research and products, the placing of many economic eggs in many economic baskets. It means large size, wide operations, formidable investment, broad vision, and constant venture into unexplored fields.

The road of security in the American chemical industry is necessarily the road of discovery; and nobody yet has been able to monopolize, more than momentarily, that broad but fickle common right of way.

Again to quote Lammot du Pont:

"The first principle of the business program is lower costs for all materials that enter into living.

"The second principle is maintenance of high wages.

"The third principle is improvement in the quality and usefulness of all existing goods, and the development of wholly new materials and new goods through scientific research and invention.

"The fourth principle is the creation of new tools, of new equipment, and new power facilities—that is, new capital added to that which we now have—in order to provide full employment not only to the present generation but also to increasing future generations.

"Simple and tried in its essentials, this program has evolved out of a century and a half of experience by practical men in the world's most highly industrialized nation. It is the product of democracy, and, being that, it places the bulk of responsibility for social betterment squarely on the shoulders of free private enterprise, where it belongs. And free enterprise accepts that responsibility, asking only a fair opportunity to work, a fair opportunity to earn a profit, and its world-old right to grow through the reinvestment of a fair part of its earnings in new enterprise."

Mr. du Pont was speaking, here, of business in general.

Nonetheless, he summed up the philosophy that ruled the Du Pont Company in 1940; and, if you should call on him in his office in Wilmington, he would tell you—as his successor to the Du Pont presidency, Mr. Carpenter, would also—it is the philosophy that has governed the Du Ponts from the beginning. Too, Mr. du Pont might produce a small chart that is a graphic record of the results of the company's ways of business in recent times. Certain lines of the chart, as they are prolonged from year to year, are the running evidence of whether or not the company has kept the ancestral faith.

Beginning with 1932, on that chart is a line identifying taxes. It darts sharply upward, ever upward. A second line indicating the course of raw material costs also ascends, but more leisurely. Then shown are three lines representing, in turn, Du Pont wages, sales, and operating profits per-dollar-of-sales. Each exhibits marked upward tendencies until 1937, after which, of the three, only the wage line resists the drop of 1938 in general business.

Mr. du Pont and Mr. Carpenter probably would pass over that wage line, leaving you to catch its significance unprompted; but they would not fail to call your attention to still another line, the bottommost on the chart. That line is labeled *"Du Pont Sales Price Index,"* meaning the composite price of all the thousands of products the company sells. For the year of 1928, it starts with 100, but thereafter, without more than minor interruptions, the line continues down, ever down, until at the end of 1939 it stands at about 67, a decline of 33 per cent.

Mr. du Pont might repeat that new products, developed largely since 1928, represented 40 per cent of the company's total sales volume in 1939, and gave direct employment to 18,000 men and women on Du Pont wage and salary rolls. Those 18,000 "jobs," he will emphasize, are new "jobs." Venture capital created them, and savings created the venture capital.

"When Adam left the Garden of Eden, he had no tools, no capital," Mr. du Pont said in an address at Boston, in 1938. "He was told, 'In the sweat of thy face shalt thou eat bread.' Through the centuries since, mankind has tempered the sweat and steadily improved its bread, which today we call the standard of living, by applying one clear rule—reinvestment of savings in the building of more and better implements.

"More capital—the venture capital that blazes new trails, and that is distinct from credit, which shies from venture—has been the requisite means toward all business progress since civilization began.

"Such capital can be built only out of the surplus that is not consumed by current needs, and from the savings of people who can afford the delays and possible losses of experimentation. Confiscate those savings, by taxation or otherwise, and in effect you confiscate progress itself!"

Mr. du Pont really believes in that doctrine. Du Pont management believes in it. It is the most basic of all Du Pont policies.

Index

Pont, 205-6; attitude toward stock-purchase bonus plan, 208; reports to Coleman, on sales to Allies (1915), 209; rejects idea of sale of company to Germans, 210; 211; heads syndicate buying Coleman's stock, 212; re-elected president, 213; defendant in suit, 218; vindicated, 219; 226; telegram from Secretary Baker, 236; proposes new contract for building powder plant for U. S., 241; letter on urgency for Old Hickory, 242-43; 261; report for 1918, 270-71; recommends new Executive Committee, 272; retires from presidency, 273, 364; 274, 276, 283; retires as chairman of board, 363; on Finance Committee, 368; 372

Dupont, Samuel, watchmaker, 3

du Pont, Admiral Samuel Francis, birth, 42; Civil War record, 101-2; death, 102

du Pont, Sophie Madeleine (daughter of E. I. du Pont), 62, 68

du Pont, Sophie Madeleine Dalmas, marries Irénée du Pont, 12; 13; gives birth to daughter, 15; at Bois-des-Fossés, 16, 18; 20, 22, 32, 35, 44, 57, 59; illness, 61; death, 62

du Pont, Thomas Coleman, his aid solicited by Alfred I. du Pont, 170; genius for organization, 170-71; previous career, 171 ff.; conference with Alfred and Pierre, 173; negotiates purchase of company, 174; president of new corporation, 175; in step with industrial trends, 180-82; purchases Laflin & Rand Company, 178, 180-82; moves offices to Equitable Building, Wilmington, 183; on Executive Committee, 189; develops leaders, 190; relations with Waddell, 192-93; 195, 200; satisfaction in company, 202; new adventures (hotels, etc.), 203; ill health and proffered resignation as president, 203-4; his bonus plan for company executives; 205 ff.; stock-sale bonus plan, 206 ff., ill-health and operations, 206; questions Alfred's motives, 209-10; sells stock to syndicate, 212; resigns presidency, 213; 214; testimony in suit,

218; his hotels and highway system, 219; U. S. Senator, 219-20; death, 220; vision, 220; 226, 261, 364, 372

du Pont, Victor Marie, description, 6; in feudal ceremonial, 6-9; travels, 10; secretary to de Moustier in U. S., 12-13; described by Mlle. de Pelleport, 17; marries her, 17; in diplomatic service of France in U. S., 17, 18; returns to France, 21; seeks to discourage colonizing company, 22; 23; 25; buys property at Alexandria, Va., and becomes naturalized, 26, 27; revisits France, 31; in commission business in New York, 33, 37; receives powder from brother's mills, 39; advertises its sale in New York, 40; bankrupt, 42; activity in Genesee Valley, 42; family, 42; brother loans credit to, 47; starts woolen mill on Brandywine, 47; 51, 57, 60; death, 61; 101, 139

du Pont, Victorine, birth, 15; 23; 32; marriage to Ferdinand Bauduy, 52; 62, 68

du Pont, William, partner in company, 124; ally of Lammot du Pont, 132; construction work at Repauno plant; 132; officer Repauno Company, 133; aids Lammot in developing wheel-mixer, 137; president Repauno Company, 139; moves offices to Wilmington, 139; in conference (1889), 142; retires from business, 146; on Finance Committee, 207; attitude regarding stock-purchase, 208-10, 212, 213

du Pont, William, Jr., 369

du Pont, William K., at Carney's Point, 190; note on, 190 n.

Du Pont American Industries, government agent, 247

Du Pont Building, 200

Du Pont Circle, Washington, 102

Du Pont Company, *see* Du Pont de Nemours Père, Fils et Compagnie; E. I. du Pont de Nemours & Co.; E. I. du Pont de Nemours Powder Company; E. I. du Pont de Nemours Company of Delaware; Du Pont, Bauduy & Company

du Pont de Nemours, Pierre Samuel, Inspector General of Commerce, his

INDEX

Repauno Chemical Co. (name from Repaupo Creek), plant built, 132; incorporated, 133; 134, 135, 146 ff., 178, 184, 191, 198, 216; new plant erected, 226

Republican Calendar, French, 15

Richmond, Va., Du Pont plants at, 308, 314

Richter, Wm., general manager, Fabbrics and Finishes Department, 368

Riker, John L., 177

Rintelin, Capt. Fritz von, seeks to buy Du Pont Company for German interests, 210

Robespierre, Maximilien de, French Revolutionist, 16; executed, 16

Robinson, Edmund G., general manager Organic Chemicals Department, 368-69

Rochester & Pittsburg Coal & Iron Company, 147

Rodman, Capt. Thomas J., invents Rodman gun, 88; invents pressure gauge, 89; with Lammot du Pont develops Mammoth powder, 89; 90; 98

Rodney, Cæsar A., U. S. Senator, 59

Roessler & Hasslacher Chemical Co., acquired by Du Pont, 277, 349

Rousseau, 26

Royalist prisoners, French massacre of, 14

Rubber, chemicals of Du Pont Company, 292-93; scarcity of, in Germany (1914-1918), 324-26; prices of (1922-1925), 326; Du Pont research for rubber replacement, 326 ff., 343, 365

Rue de la Corderie, Paris, Du Pont apartment in, 14

Russia, 80

Saltpeter, U. S. Government's, 50; prices of Indian, 50; prices of Mammoth Cave saltpeter, 51; chief component of black powder, 81; Indian and Peruvian compared, 81; Indian exhausted, 83; 94-95; scarcity of, embarrasses Union, 96-97; deposits in South, 96; supply problems solved by chemistry, 98, 99; 102; 128; 134

San Jacinto, U. S. S. S., 95, 96; takes Confederate commissioners from *Trent*, 95; 96

Santo Domingo, French troops in, 33, 37

Sargent, Attorney-General John G., absolves Du Pont Company of "fraud or crime," 250

Scarlett, James, chief of Government counsel in suit against Du Pont Company, 195

Schoenbein, Prof. C. F., invents guncotton, 107; 153, 155; 157

Scott, F. A., chairman Munitions Board (1917), 231

Seaford, Del., Du Pont nylon yarn plant at, 360

Seed disinfectants, Du Pont project, 278

Serajevo, assassinations, 205

Seven Pines, Va., Du Pont Company operates plant at, 247

Seward, Wm. H., 97

Sherman Anti-Trust Act, 182, 193, 195

Simonton, Wm. A., 271

Slidell, John, Confederate commissioner, taken from *Trent*, 95; surrendered, 97

Sloan, Alfred P., Jr., in charge General Motors, 276, 369

Smokeless powder, 157, 181, 198; Allied orders for (1915), 208, 209; Du Pont Department, 226

Soaps, 347-48

Sobrero, Prof. A., invents nitroglycerin, 107, 108; 111, 113

"Soda powder," invented by Lammot du Pont, 82; demand for, 82; description, 82-83; 89

South Carolina, resists Federal authority (1833), 63; powder mill in, in 1861, 90

Spanish-American War, 161, 188

Sparre, Fin, Director, Development Department, 369

Spruance, William C., Jr., on Executive Committee, 272, 282-83; as vice-president negotiates for "Cellophane," 311

Staël, Mme. de, 31

Standard Oil, 179

Stephenson, George, his *Rocket*, 69

Stevenson Act, to limit rubber production (1922), 326; repeal, 326